AF326815

Injection Technologies for the Repair of Damaged Concrete Structures

V.V. Panasyuk • V.I. Marukha • V.P. Sylovanyuk

Injection Technologies for the Repair of Damaged Concrete Structures

 Springer

V.V. Panasyuk
National Academy of Science of Ukraine
Karpenko Phisico-Mechanical Institute
Lviv
Ukraine

V.P. Sylovanyuk
National Academy of Science of Ukraine
Karpenko Phisico-Mechanical Institute
Lviv
Ukraine

V.I. Marukha
National Academy of Sciences of Ukraine
State Enterprise "Engineering Center
  (Techno-Resurs)"
Lviv
Ukraine

ISBN 978-94-007-7907-5   ISBN 978-94-007-7908-2 (eBook)
DOI 10.1007/978-94-007-7908-2
Springer Dordrecht Heidelberg New York London

Library of Congress Control Number: 2013955493

© Springer Science+Business Media Dordrecht 2014
This work is subject to copyright. All rights are reserved by the Publisher, whether the whole or part of the material is concerned, specifically the rights of translation, reprinting, reuse of illustrations, recitation, broadcasting, reproduction on microfilms or in any other physical way, and transmission or information storage and retrieval, electronic adaptation, computer software, or by similar or dissimilar methodology now known or hereafter developed. Exempted from this legal reservation are brief excerpts in connection with reviews or scholarly analysis or material supplied specifically for the purpose of being entered and executed on a computer system, for exclusive use by the purchaser of the work. Duplication of this publication or parts thereof is permitted only under the provisions of the Copyright Law of the Publisher's location, in its current version, and permission for use must always be obtained from Springer. Permissions for use may be obtained through RightsLink at the Copyright Clearance Center. Violations are liable to prosecution under the respective Copyright Law.
The use of general descriptive names, registered names, trademarks, service marks, etc. in this publication does not imply, even in the absence of a specific statement, that such names are exempt from the relevant protective laws and regulations and therefore free for general use.
While the advice and information in this book are believed to be true and accurate at the date of publication, neither the authors nor the editors nor the publisher can accept any legal responsibility for any errors or omissions that may be made. The publisher makes no warranty, express or implied, with respect to the material contained herein.

Printed on acid-free paper

Springer is part of Springer Science+Business Media (www.springer.com)

Preface

Concretes and reinforced concretes find wide application in the construction of industrial structures such as bridges, tunnels, large reservoirs, collectors, and docks, as well as residential buildings. During use, such structures are subject to degradation, damages and local fracturing. This causes loss of serviceability of the damaged structural elements and danger of fracture of the overall structure. In order for safe use of an impaired structure, it is necessary, on the one hand, to evaluate its residual serviceability, that is, the remaining safe service life, and on the other hand, to apply effective technologies for the renewal of serviceability by "healing" the impaired or damaged elements.

This book summarizes and analyses the most important achievements of science and engineering concerning the service factors that cause damage to concrete and reinforced concrete structures; methods for assessing their strength and life, especially those that are based on modern concepts of the fracture mechanics of materials, and basic approaches to predicting the residual life of structures of long-term operation.

Concrete injection technologies for the restoration of serviceability of impaired concrete and reinforced concrete structures by means of healing cavities, cracks, fissures, corrosion injuries, etc., were given special attention. Case studies of implementation of the above technologies for the restoration of integrity and extension of service life include concrete and reinforced concrete structures such as atomic power plants (APP), underground railway, bridges, seaports, historical relics and others.

The outlined principles of structural material strength assessment, injection technologies for the restoration of serviceability of impaired concrete structural elements and presented case studies will be useful for building specialists, lecturers, post-graduate students and students of high schools specializing in building.

The authors of this book would like to express deep gratitude to the employees of the Engineering Centre "Techno-Resurs" (Kiev, Ukraine), B. Y. Henega, I. P. Hnyp, and Y. A. Serednyts'kyi, for providing the informational materials on the application of injection technologies for renewal of damaged concrete and reinforced concrete structures. The authors are grateful to their colleagues from Karpenko Physico-Mechanical Institute N.A.S. Ukraine, M. A. Ivantyshyn, N. V. Onyshchak, and R. Y. Yukhym, for assistance in preparation of this book.

Contents

List of Abbreviations

a, b	semi-axes of an elliptical plane crack
d	distance between crack centers
E, E_1	Young moduli of matrix and filler, respectively
$E(k)$	complete elliptic integral of the second kind
$2h$	solidified injection material layer thickness
$H(\zeta)$	Heaviside function
K_I, K_{II}, K_{III}	stress intensity factors (SIF) for cleavage crack, transverse shear crack, and longitudinal shear crack, respectively
$K_{IC}, K_{IIC}, K_{IIIC}$	fracture toughness characteristics of material (critical SIF values)
$K(k)$	complete elliptic integral of the first kind
$2l$	crack length
p, q	applied tensile or shear load intensity coefficients, respectively
p_c, q_c	critical applied tensile or shear load, respectively
R_{bt}	ultimate tensile strength (breaking resistance)
R_b	ultimate compressive strength
R_{BT}	ultimate fatigue strength
t	specimen thickness
u_x, u_y, u_z	components of displacement vector $\vec{u}$
β	form factor of cavity or crack
γ	specific fracture energy
$\delta_I, \delta_{II}, \delta_{III}$	crack opening displacement (COD) for cleavage, transverse shear, or longitudinal shear mode, respectively
δ_{IC}	critical crack opening displacement
ε_c	ultimate tensile strain
ε_T	ultimate plastic strain
ν, ν_1	Poisson's ratio of matrix and filler, respectively
Σ_b	ultimate strength
σ_{ij}	stress tensor components
$\sigma_T, \sigma_{0.2}$	conventional and physical yield strength, respectively
τ_c	ultimate shear strength

Chapter 1
Introduction

Abstract This book presents the main physical and mechanical characteristics of concrete and reinforced concrete materials as well as evaluation methods and typical damage that can be expected over long-term use.

The authors pay special attention to the description and use of up-to-date injection technologies for strengthening the damaged structural elements of a long-term operation; they present recommendations for choice of materials for the implementation of such technologies. The book contains criteria for injection materials and examples of compliant polymer materials.

The methods and means for technical diagnostics of damaged concrete and reinforced concrete constructions of long-term operation are given a significant place. Damage mechanisms are considered. The information presented includes data concerning equipment and appliances used in injection renewal works. The case studies provided illustrate the removal of damage and the extension of service life for certain renewed objects of long-term operation.

The intended audience is comprised of building engineers, scientists, lecturers, and high school students specializing in construction.

This book contains 192 illustrations, 26 tables, and 211 references.

Structures that have been in operation for over 50 years and need repairing because of assorted damages make up the majority of residential and other concrete and reinforced concrete constructions in Ukraine. Many unique structures of historical and architectural importance exist in a state of potential accident. Degradation of service characteristics of the reinforced-concrete materials in APP structures, mine shafts, collectors, and hydraulic works have given birth to the necessity of serviceability renewal, repair, and the restoration of carrying ability in such structural elements. Under this circumstance the development of engineering methods for the diagnostics, serviceability and operation reliability renewal of buildings in long-term operation, and especially the development of effective technologies for practice, are very important from both economic and technical viewpoints.

Most urgent scientific and engineering aims include the following:

- Reliability and life improvement of building concrete and reinforced concrete structures;
- Development of technologies and necessary materials for the renewal of impaired elements in such structures in order to extend the reliable operation life, and

V. V. Panasyuk et al., *Injection Technologies for the Repair of Damaged Concrete Structures*, DOI 10.1007/978-94-007-7908-2_1,

© Springer Science+Business Media Dordrecht 2014

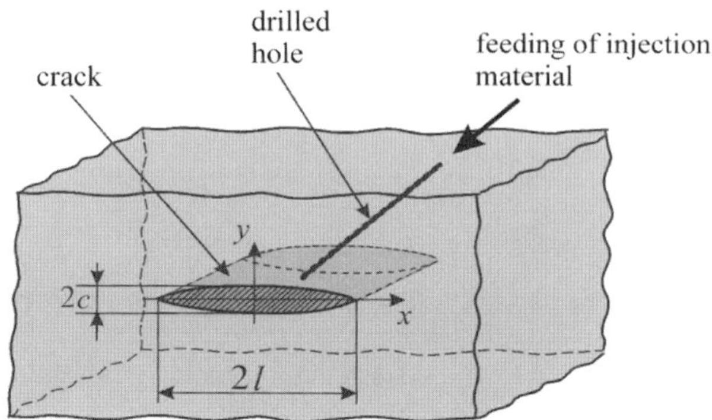

Fig. 1.1 Sketch of the healing of a cracked structural element using concrete injection

- Adaptation of scientifically sound concepts for evaluation of the structural strength and life of concrete and reinforced concrete, as well as concepts for optimization of renewal technologies in the case of degradation during long-term operation

Domestic and foreign practices use different technological approaches to the renewal of impaired structures and constructions of long-term operation. Depending on the type of structure, nature of impairment, conditions of use, etc., different repair methods and technologies have proven most effective. The existing large-scale structures suffering impairment and undergoing renewal using different technologies can be put into the following categories:

- Dams, canals, locks, and other hydraulic facilities;
- Residential houses;
- Underground railway and motor transport tunnels, bridges, overpasses and underpasses;
- Industrial and municipal sewage water collectors of large diameter;
- Wharfs and other constructions in sea and river ports;
- Building basements in watered, saline or microbiologically active soils

Concrete injection into the defect zone is one of most effective methods of healing the damaged structural elements (see Fig. 1.1). This method consists of introducing a liquid material, usually a polymer-based material, into the damaged zone (containing cracks, voids, crumbling, lamination, etc.) of a concrete or reinforced concrete structure. Such liquids are able to form strong adhesive bonds with a concrete matrix after solidification. In recent years, these technologies have been widely used in both domestic and foreign practices for the integrity, strength and restoration of serviceability of many structures.

Among the wide variety of injection materials based on viscous liquid compositions (silicate, silicon-organic, acrylic, phenol formaldehyde, etc.), polyurethane based mixtures are the materials of choice due to their relatively simple preparation and pumping technologies.

Polyurethane elastomers possess many high, sometimes unique, technical and service properties. These properties, combined with a rather good adhesion to the concrete matrix in cracks and damages, provide reliable operation of the renewed

Fig. 1.2 Fragment of a structure with cracks and traces of corrosive environment penetration

structures under continuous action of force and corrosive environments, as well as other factors that cause the concrete to degrade and break. In particular, low viscosity of polyurethane injection compositions pumped into a concrete mass under pressure $1 \div 25$ MPa makes it possible not only to fill in the internal cavities and damage but also completely to remove the corrosion and mechanical injuries. Moreover, the solidified polyurethane elastomers can block the propagation of cracks and deter the penetration of water containing dissolved corrosion and active microbiological impurities.

Thereby, polyurethanes eliminate one of the most widely spread factors of corrosion and mechanical injury of a concrete matrix. Besides, strong adhesive bonds with the concrete matrix create an essentially continuous composite material in place of the defect zone.

The main performance characteristics causing damage and, in some cases, fracture of concrete structures and constructions over long-term operation are, first of all, the microstructure and composition of the concrete, mechanical loads, corrosive environments, temperature and/or hydraulic pressure variations, etc. The complicated composition of concrete causes the occurrence of numerous microstructural heterogeneities, capillaries, open pores and internal cavities on both the surface and in the bulk of the material that transform into open or sharp-notched voids, micro- and macrocracks, laminations, etc., during use. These defects drastically enhance the penetration of corrosive constituents from the environment into the materials of building structures, thereby accelerating degradation and fracture of such structures (Fig. 1.2).

The following outer environments are especially dangerous for concretes: aqueous solutions of acids, alkali, salts, organic compounds and others. In many cases, microorganisms (soil bacteria, fungi, lichens) enhance the destructive effect of chemical agents. Favorable conditions for the microbiological corrosion of concretes and other building materials appear when microorganisms grow on the surfaces and, especially, in defects caused by aqueous environments containing cations and anions of variable valence and organic food substrates. The simultaneous effect of chemical and microbiological factors is especially dangerous. Such conditions are typical, as a rule, for underground concrete structures.

Fig. 1.3 Seaport berth
damaged by waves and
corrosive environment

Use of concrete constructions and structures in seaports, coastal territories and other environments of corrosion and active microbiology is especially harmful due to the permanent damaging effect of corrosive environments.

In addition to the environmental effect, concrete structures undergo continuous static and cyclic mechanical loads during operation. Such loads include the stresses created during construction or mounting and those generated during use due to basement or foundation skewing, as well as sea wave slaps, vibrations from transport movement, etc. (Fig. 1.3)

In practice, concrete structures experience the simultaneous effects of corrosively and mechanically harmful environments. One of the most dangerous mechanisms of damage and fracture of concrete and reinforced concrete structures in this case is the formation and propagation of crack-like defects. These defects can significantly decrease the strength of the structure and cause loss of element serviceability.

This book, "Concrete Injection Renewal Technologies of Impaired Long-Term Structures," consists of seven chapters.

Chapter 2 (authors V. I. Marukha, V. V. Panasyuk, V. P. Sylovanyuk) describes the general characteristics, classification, and grades of concretes and reinforced concretes. Since concretes are heterogeneous materials, the influence of microstructure on strength and deformation properties is significant. Typical diagrams of deformation and failure for various concretes are included. The authors present a mathematical model of the compressive and tensile strength of concretes. This chapter contains an outline of the basic characteristics of crack growth resistance (fracture toughness) and evaluation methods for concrete. Tables in this chapter present the service characteristics of different concrete grades.

Chapter 3 (authors V. I. Marukha, V. V. Panasyuk) considers examples of concrete microstructure degradation under the influence of mechanical loading, corrosive environments, and long-term operation. Characteristics of concrete operation and fracture under short-term static loading, long-term static loading and cyclic loads are included. Damaging mechanisms of concrete structural elements are given. Corrosion injuries in concretes under the influence of different factors are classified. This classification defines three types of concrete corrosion as well as gas and biological corrosion. In addition, the authors include reinforcement corrosion in reinforced concrete under static and cyclic loading as a separate class.

Chapter 4 (author V. I. Marukha) contains information about equipment and devices that are used for the implementation of injection renewal on impaired objects. In particular, the author provides a functional chart of the mobile truck-mounted diagnostic and restoration complex, which was developed and designed at Karpenko Physico-Mechanical Institute NASU and State Engineering Center "Techno-Resurs" NASU (Kiev, Ukraine). This complex comprises devices for damage diagnostics; a computer system for information recording and processing; remotely controlled sensors for checking the status of building constructions (underground collectors, pipelines, houses, etc.); injection equipment set; and auxiliary equipment (*see Fig. 1 in the appendix*). The chapter contains data on similar up-to-date foreign equipment as well (Fig. 1.4).

Chapter 5 (author V. I. Marukha) presents general information on the physical and mechanical characteristics of materials for injection renewal technologies. The main technological requirements for the injection materials include high flow ability ensuring maximum possible filling in of the defects (cavities, laminations, cracks); sufficient adhesive cohesion for the restoration of structure serviceability; quick solidification and hardening, etc. Approaches for evaluating these characteristics and optimization of the technological parameters for injection are then proposed.

Chapter 6 (authors V. V. Panasyuk, V. P. Sylovanyuk) describes analytical models and solutions for specific problems concerning the strength of deformed bodies with defects filled with injection materials. The obtained solutions are the theoretical basis for estimations of service life for structural elements after renewal by injection technologies.

For this purpose, a mathematical model of a cracked material healed with injection technologies was developed. The authors analyzed the model in both 2D and 3D formulations. This chapter presents the applicability limits of results obtained in two-dimensional approximation. Investigation of the effects of crack wedging by injection mixtures shows that such effects are significant since they can lead to growth of the initial crack-like defects under certain conditions.

The extent to which the defects can be filled with an injection material is the important parameter of the technology under consideration. Complete defect filling is often very difficult to attain in practice, for many reasons. Therefore, it is important to develop approaches for evaluating the influence of incomplete defect filling on the effectiveness of the renewal of damaged structural elements.

The problem of injection into a damaged body containing a system of mutually interacting cracks is also considered. The authors consider injection into a system of two cracks in detail. They studied the effectiveness of the restoration of strength for the case of cylindrical structural elements. The solution for transverse compression of a cylindrical element along the planar defect is included. Such specimen configuration is widely used in the testing of brittle materials for strength and fracture toughness. The convenience of such a configuration is that it requires no special equipment for experiments except for a compression machine. The basic experimental investigations necessary for the optimization of injection technologies have been performed using this scheme alone.

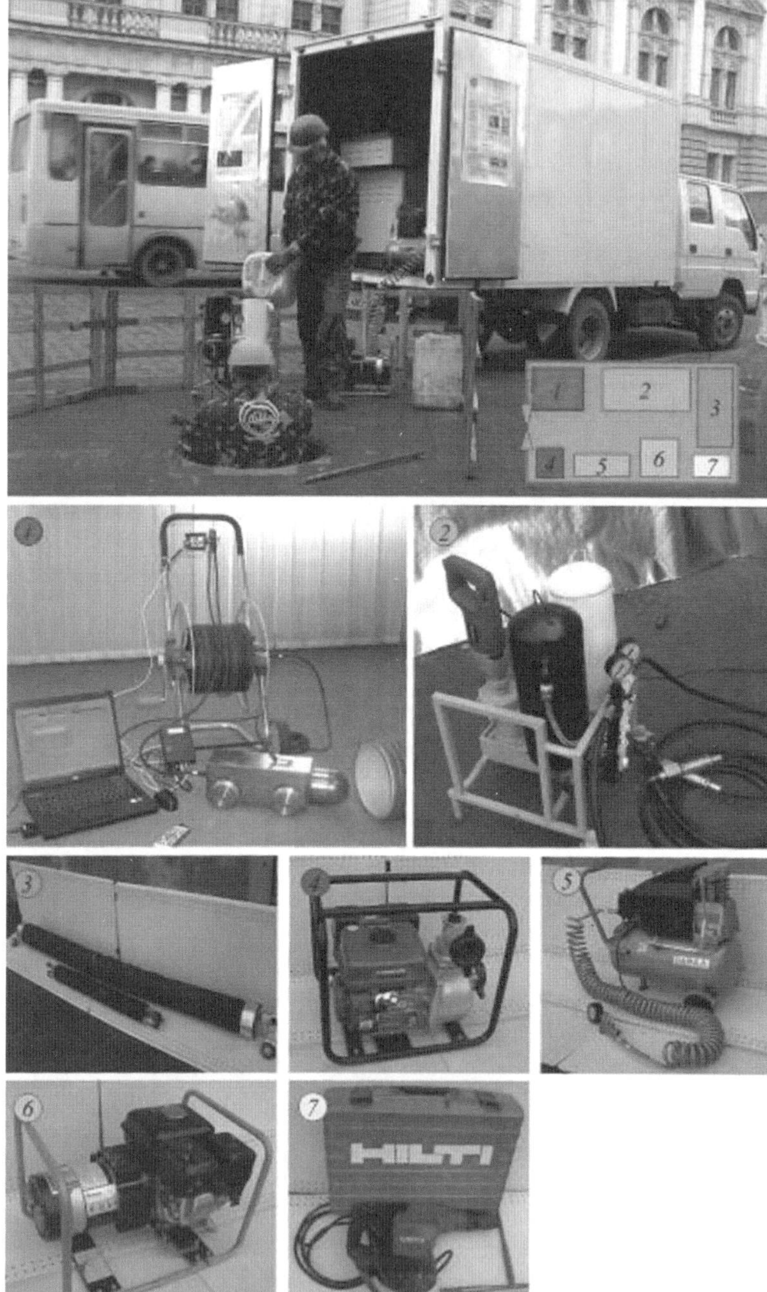

Fig. 1.4 Mobile diagnostic and restoration complex: arrangement of equipment packages inside the van

Existing in concrete materials, voids and pores play a significant role in the formation of cracking defects. The concentration of tensile stress arises around such defects even under compression, which facilitates the initiation and propagation of cracks in the loaded concrete. Filling the cavities with injection material significantly strengthens the concrete, as indicated by the results of theoretical and experimental studies.

The present work establishes, in particular, the injection parameters at which a fatigue crack will stop. The work also determines a number of cycles for each fatigue-loading regime, during which the crack will preserve its length without growth,

The chapter also contains results of the study of the strength of the concrete–polyurethane joint adhesive, strength of concrete specimens with crack-like (filled with injection materials or empty) concentrators, increase of Young's module of injected materials, etc., concerning the restoration of high serviceability of renewed structural elements.

The adhesive strength of injection material cohesion with the concrete matrix is one of the basic criteria for the selection of injection material. Preferably, adhesive strength must be higher than the strength of the concrete matrix. The authors have tested six types of injection materials. The polyurethane composition "EKOPUR HW" served as the base agent for strengthening specimens with flaw-like stress concentrators. Adhesion strength tests consisting of tearing off the steel disks glued to the concrete surface with these materials revealed that fractures transected the concrete body in all cases, which means that the adhesive strength of the joint is higher than that of the concrete matrix.

Strength test schemes for injection-renewed structural elements with artificial defects were as follows: (a) transverse compression of a cylindrical specimen with or without flaw-like concentrator (filled or empty) along the concentrator plane; (b) bending of a prismatic specimen with or without concentrator (filled or empty); (c) compression of a prismatic specimen with or without holes (filled or empty), see Fig. 1.5.

Test results confirmed a good conformity between the experimental and theoretical data on the strength of the injection-renewed concrete specimens with model crack-like defects.

Chapter 7 (authors V. I. Marukha, V. P. Sylovanyuk) describes the main methods of technical diagnostics of the state of long-term service constructions based, as a rule, on non-destructive inspection techniques. The main defects in concrete and reinforced concrete materials include cracks, voids, laminations, corrosion injuries, etc. Many methods exist to reveal such defects, with the most widely used being visual, optical, acoustic, thermographic, radiation, and penetrating agent measurements.

Each method uses certain physical processes and phenomena and consists of measuring the studied structure's response to a certain applied physical factor, in particular, electromagnetic, thermal, mechanical, radiation or other fields. The response of the measured structure is subject to comparison with the similar response of a conditionally perfect (defectless) structure under the same circumstance.

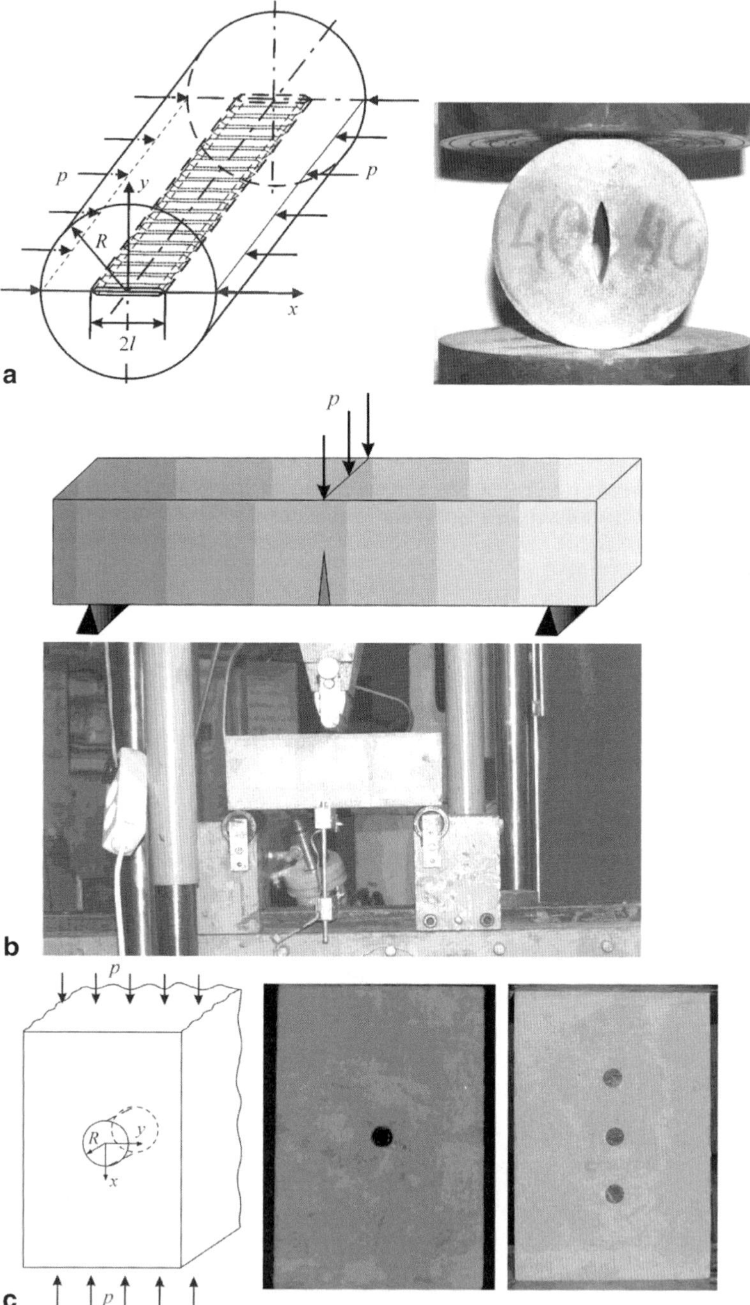

Fig. 1.5 Scheme of loading and fracture of specimens with concentrators: **a** transverse compression of a cylindrical specimen along the flaw plane; **b** three-point bending of a prismatic specimen; **c** compression of prismatic specimens with holes

Hydroelectric power station   Architectural monuments

Underground collectors   Bridge footing

Residential houses   Underground railway

Fig. 1.6 Examples of structures in long-term operation as objects for the application of injection strengthening technologies

The revealed difference, that is, the deviation of a parameter from the reference value measured in the perfect structure, characterizes the defect occurring in a real material. Subsequent additional investigations allow for identifying the defect (geometrical form, dimensions, nature, etc.) using respective measuring techniques. The chapter contains examples of non-destructive testing and inspection techniques and devices for revealing defects in the concrete elements of building structures.

Finally, Chap. 8 (authors V. I. Marukha, V. V. Panasyuk, V. P. Sylovanyuk) acquaints the reader with case studies of long-term service constructions where injection technologies allowed for the elimination of defects and the restoration of serviceability. Such structures are as follows: underground constructions (collectors, tunnels, and railway); hydraulic works (dams, bridges); APP objects; residential houses; architectural monuments, etc. (Fig. 1.6).

The presented case studies demonstrate the high efficiency, feasibility and prospects of the up-to-date methods for the renewal of impaired structures using injection technologies.

Chapter 2
General Characteristics of Concretes and Reinforced Concretes

Abstract The Chapter presents general characteristics of concretes and reinforced concretes, their classification, grading and designations. Definition of structure inhomogeneity in these materials and its effect on service characteristics are given. The Chapter contains typical diagrams of concrete deformation and fracture as well as nominal strength values for different concrete grades in tables. Besides strength, other important mechanical characteristics, such as fracture toughness, are of no less importance in concretes. The authors provide a summary of existing fracture toughness criteria and methods for experimental fracture toughness determination. At present, concrete science can formulate only individual aspects of a concrete strength and deformation theory. Considering concrete as a composite material with hierarchical structure (at micro, meso, and macro levels) and analyzing these structure levels, the authors propose a mathematical model enabling the establishment of quantitative relations between breaking (tensile) and compressive strength values for a porous material such as concrete.

The Chapter contains classification and mechanical properties of concrete reinforcement as well.

Concretes are conglomerates formed through the solidification of a mix of cement solution, water, fillers, and modifying additions, if needed. A wide diversity of binding cement materials and concrete fillers (aggregates) as well as physical and mechanical characteristics of the concrete components significantly complicates the development of generalized microstructure models and strength theories for concrete [1]–[6].

Reinforced concrete is concrete enhanced with steel rods. While the elastic moduli of the cement stone and aggregate differ significantly but within one order of magnitude, the elastic modulus of a steel is nearly a full order higher than the integrated elastic modulus of concrete as a whole.

Concrete intended for use in reinforced structures must possess the following mandatory pre-determined physical and mechanical properties: high strength; good adhesion to reinforcement; sufficient density (moisture impermeability) for protection of reinforcement from corrosion, etc.

Depending on the intended application and operation conditions, the concrete must comply with the following special requirements: resistance to corrosive environments (e.g., reservoirs in the chemical industry); heat resistance (including long-term operation at high temperatures); high fracture toughness (i.e., resistance to crack propagation), etc.

V. V. Panasyuk et al., *Injection Technologies for the Repair of Damaged Concrete Structures,* DOI 10.1007/978-94-007-7908-2_2,
© Springer Science+Business Media Dordrecht 2014

As adopted in the practice of engineering, the short names of concrete grades intended for use in reinforced structures are as follows:

a. *Heavy concrete* is concrete with a dense microstructure containing heavy aggregates bonded at the cement mortar solidification.
b. *Fine concrete* is concrete with a dense microstructure containing fine aggregates bonded at the cement mortar solidification in any ambient conditions.
c. *Lightweight concrete* is concrete with a dense microstructure containing porous aggregates bonded at the cement mortar solidification in any ambient conditions.

Heavy aggregates may consist of crushed rocks (sandstones, granite, diabase, etc.) and natural quartz sand. Porous aggregates may be natural (pearlite, shell limestone, etc.) or artificial (expanded clay, slag, etc.). Depending on the type of porous aggregate, there exists expanded-clay (ceramsite) concrete, slag concrete, pearlite concrete, etc.

Cellular concretes, or concretes with porous aggregates and medium density $1400 \, kg/m^3$ or lower, are mainly applied in shielding structures. Heavy concretes are most commonly applied for protection from harmful radiation.

To prepare concretes satisfying certain special requirements, for example, of a specified mechanical strength, the concrete should have certain components in a predetermined quantitative ratio including various cements, coarse and fine aggregates, modifying additions, etc.

The strength of a concrete depends on many factors including aggregate particle size; aggregate strength and surface condition; cement brand and proportion; water content during solidification; aggregate roughness and adhesion to the cement mortar, etc. [7].

2.1   Concrete Microstructure and Its Effect on Strength and Deformation Behavior

A concrete's microstructure has a strong effect on its strength and deformation behavior. In order to understand the mechanisms underlying this effect, let us consider the process of concrete formation. After flooding a dry mixture of aggregates and cement with water, a chemical reaction occurs between the cement minerals and water resulting in the formation of a gel being the fluid mass, which consists of cement particles and various crystalline compounds suspended in water. During agitation of the concrete mix, the gel envelops aggregate particles and gradually solidifies, while suspended crystallites combine into progressively growing crystals. The solidified gel transforms into the cement stone, binding all coarse and fine components into the monolithic solid material referred to as concrete.

In the resulting microstructure, solid aggregates occupy more than 80 % of the volume, depending on the grade of concrete. The composition of the basic cement mass (gel or cement stone) comprises both minerals formed in initial reactions of the dry cement minerals with water and compounds forming in subsequent reactions

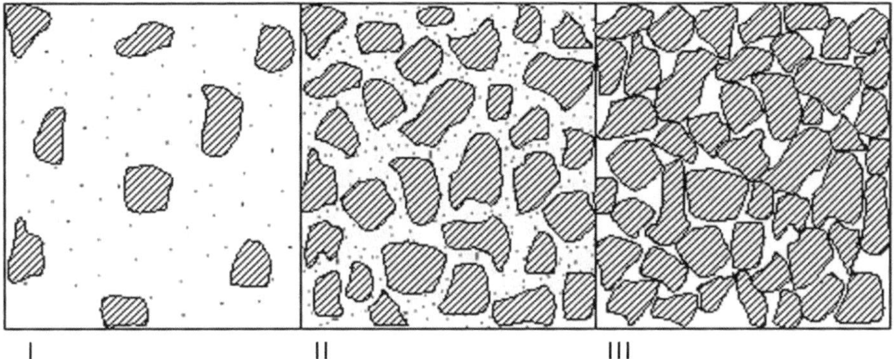

Fig. 2.1 Typical classes of concrete microstructure

between the so-formed products. The microstructure of the cement stone is generally crystalline and consists of various calcium compounds, namely, calcium hydroxide, hydrosilicates, hydroaluminates, and hydroferrites.

The characteristics of the concrete depend to a considerable extent on the microstructure density or porosity, as well as the porosity type. Using the criterion of porosity, one can define the following typical concrete classes (see Fig. 2.1):

- concrete with 'floating' aggregates (I);
- concrete with closely packed aggregates (I I);
- high-porous concrete with a deficient cement binder (I I I)

Thus, the microstructure of concretes is very heterogeneous. From a physical point of view, the concrete is a capillary-porous material containing three phases: solid, liquid, and gaseous. The cement stone, in turn, is very heterogeneous too.

Prolonged processes that continue in solid concrete due to changes in water balance, solidifying gel volume, and crystal size, impart specific elastoplastic properties to the material expressed in the deformation behavior of the concrete under action of loads and conditions of temperature-humidity.

The deformation behavior of the concrete implies its densification (compaction), swelling, and creep. In particular, concrete hardening in ambient air conditions leads to shrinkage (compaction) whereas hardening in water causes dilatation (swelling). As experiments show, the magnitude of shrinkage of a concrete depends on the following factors:

a. cement proportion and grade: lower proportion of cement volume results in greater shrinkage, highly active or alumina cements causing great shrinkage;
b. proportion of water: higher water-cement ratio (w/c) results in greater shrinkage;
c. aggregate particle size: fine sand and porous aggregates cause greater shrinkage

Concrete swelling takes place when the material hardens in humid conditions or contains special expansion agents.

Concretes prepared using certain special cement brands are non-shrinking [6].

If the element of structure of a concrete is loaded, the load will induce certain initial deformation, and thereafter, the element will change its form over time, while stresses created by the load will relax. The phenomenon of inelastic deformation under long-term loading is known as concrete creep.

2.2 Concrete Strength and Stress-strain Behavior

Concretes as structural materials are distinguishable by their high microstructure heterogeneity. Therefore, certain simplifications or approximations are required in the development of physical and mechanical models for stress state analysis and prediction of the strength and deformability of concretes. In particular, such models treat the concrete as elastic homogeneous continuum with some averaged Young modulus (E), Poisson ratio (ν), specific fracture energy (γ) (that is, energy spent in the formation of the fracture surface unit), ultimate tensile (or bending) strength (R_{bt}), and ultimate compression strength (R_b). The above characteristics are subject to experimental measurement. In the elastic continuum approximation, they describe the elastic medium and serve as the base in estimating strength and durability of the elements of concrete structures. Obviously, estimations of durability require knowledge of the above characteristics as functions of time, temperature, and diffusion processes running during longtime material operation under given conditions.

In addition to continual mechanical models, there exist analytic local models accounting for structural inhomogeneity (pores, voids, interlayers between aggregate particles, cracks and other stress concentrators) and determining resistance of the material to local crack nucleation. Crack growth from such nuclei can lead to the complete breakage of concrete and reinforced concrete structures.

Such data are necessary for the engineering practice of the renewal of the impaired structural elements in the selection of appropriate healing materials and methods and serviceability estimations.

Concretes usually demonstrate nonlinear stress-strain dependence at both tension and compression. A generalized stress-strain curve for concrete is linear at small strains and passes extremes at larger strains in areas of both tension and compression (Fig. 2.2).

Extreme stress values in the experimental plots (Fig. 2.2) represent the ultimate tensile strength (R_{bt}) and ultimate compression strength (R_b), which are measured using the testing schemes shown in Fig. 2.3.

The Young modulus is measurable in accordance with Hooke's law as a tangent of the curve angle α_0 with the strain axis within the limits of elasticity (Fig. 2.2):

$$E = k\tan\alpha_0, \tag{2.1}$$

where k is the unit conversion factor.

Since stress-strain curve is nonlinear at high (breaking) stresses (Fig. 2.2), the elastic modulus may be determined at any load value as the tangent slope at respective points of the stress-strain curve:

$$E' = k_1\tan\alpha_1. \tag{2.2}$$

Fig. 2.2 Stress-strain curve for concrete in areas of tension and compression

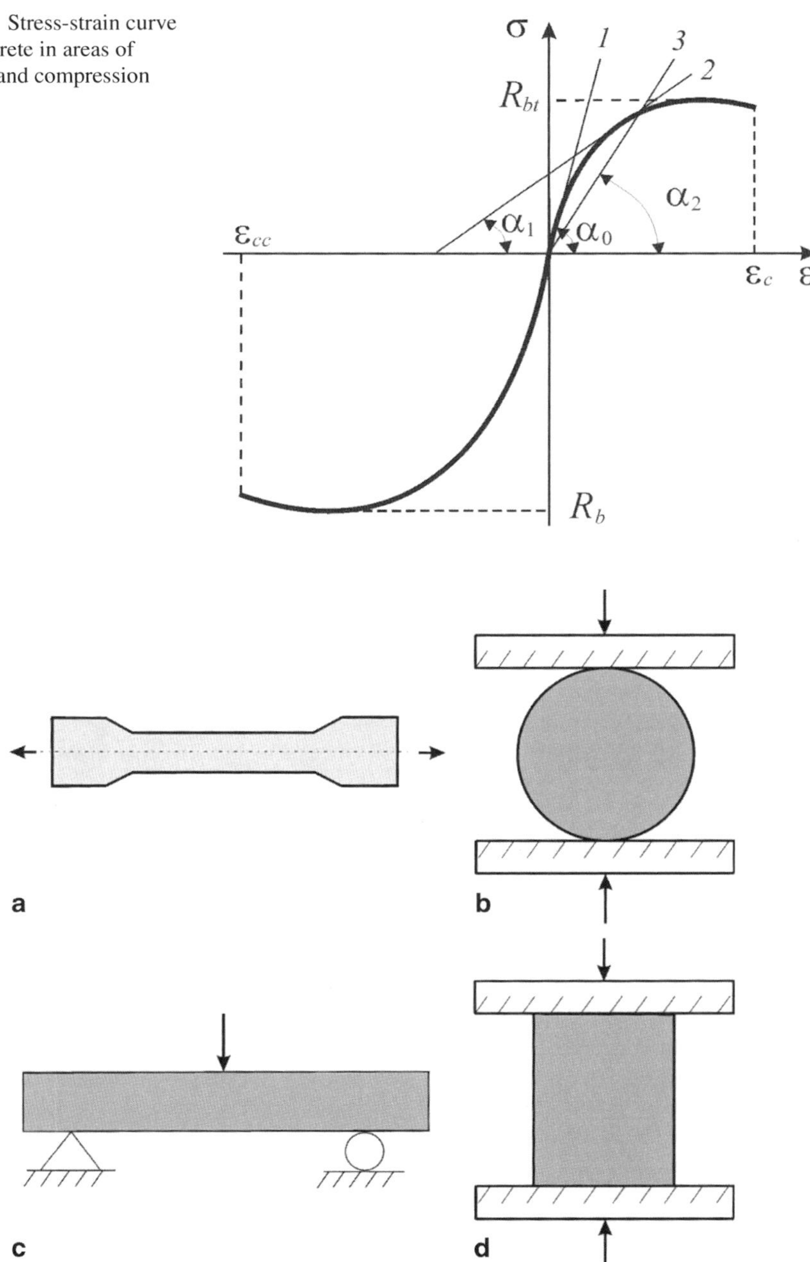

Fig. 2.3 Testing schemes for determining the tensile strength (*a–c*) and compressive strength (*d*)

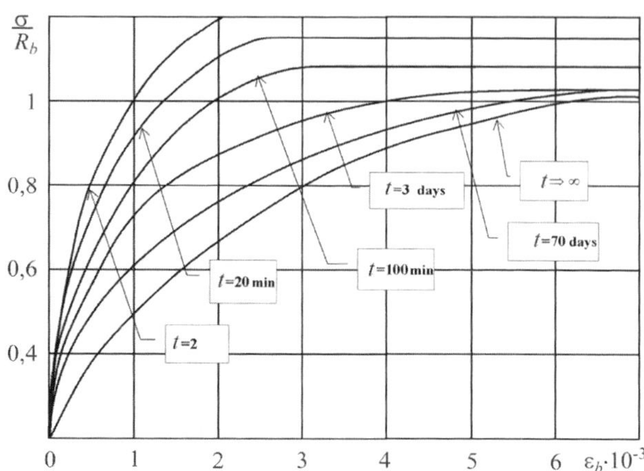

Fig. 2.4 Stress-strain curves for concrete at various loading times. [1]

An average elasticity modulus needed for calculations of the stress state is proportional to the slope of straight line *3* (Fig. 2.2) passing through the particular point at the given stress level:

$$E'' = k_2 \tan \alpha_2. \tag{2.3}$$

It is clear that so-defined elasticity moduli have variable values since angles α_1, α_2 depend on the concrete load time (Fig. 2.4).

The relationship between the elasticity modulus at tension or compression and the shear modulus is as follows:

$$G = \frac{E}{2(1 + v)}, \tag{2.4}$$

where v is the Poisson ratio. Shear modulus (G) is an important characteristic of concrete because combined stress states, for example, shear combined with tension (beams), shear combined with compression (arches), are very common in practice. At $v = 0.2$, the shear modulus amounts to $G \approx 0.4E$.

Consequently, only the initial elastic modulus determined in the initial stage of a specimen loading within limits of elastic deformation is the physical characteristic of concrete (Fig. 2.2; Eq. 2.1). Table 2.1 presents values of the elastic modulus for certain grades of concrete [1].

2.3   Compressive Strength of Concretes

Compressive strength is the most common and important characteristic of concretes. During compression, stresses in a concrete specimen concentrate either in aggregates, which have higher elastic nodulus values (are harder), or around holes and

Table 2.1 Elastic moduli of concretes at tension or compression $E_b 10^{-3}$, MPa

Concrete	Concrete class according to compressive strength (MPa)										
	B12.5	B15	B20	B25	B30	B35	B40	B45	B50	B55	B60
Heavy, natural hardening	21	23	27	30	32.5	34.5	36	37.5	39	39.5	40
Heavy, heat treated	19	20.5	24	27	29	31	32.5	34	35	35.5	36
Fine type A, natural hardening	17.5	19.5	22	24	26	27.5	28.5	–	–	–	–
Fine type A, heat treated	15.5	17	20	21.5	23	24	24.5	–	–	–	–
Fine type B, natural hardening	15.5	17	20	21.5	23	–	–	–	–	–	–
Fine type B, heat treated	14.5	15.5	17.5	19	20.5	–	–	–	–	–	–
Fine type C	–	16.5	18	19.5	21	22	23	23.5	24	24.5	25
Lightweight, graded by density:											
1400	11	11.5	12.5	13.5	14.5	–	–	–	–	–	–
1800	14	15	16.5	18	19	20	20.5	–	–	–	–

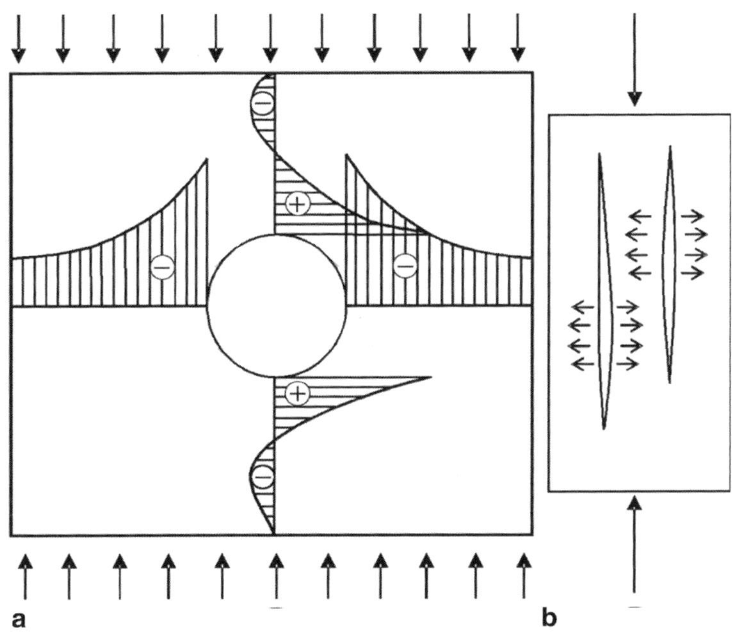

Fig. 2.5 Stress state pattern near micropores (**a**) and genesis of cleavage cracks in a concrete speci-men under compression (**b**) [8]. The symbol ⊕ here depicts tensile stresses, symbol ⊖ compressive stresses

cracks. For this reason, a complicated stress state arises in the mechanically inho-mogeneous conglomerate (Fig. 2.5) including both compressive stresses and tensile stresses, especially near cavities or voids. Under a compressive loading mode, the primary cracks parallel to the compression axis open first, becoming cleavage cracks (Fig. 2.5b).

Table 2.2 Values of the size correction factor β for compressive strength determination of concrete. [8]

Shape of specimen	Cube					Cylinder			
Size of specimen (cube edge or cylinder diameter), cm	7	10	15	20	30	7×14	10×20	15×30	20×40
β value		0.85	0.91	1.0	1.05	1.10	1.16	1.16	1.20	1.24

Thus, tensile stresses adjacent to the compressive ones originate in a concrete specimen under compression around defects like voids or cracks, which, therefore, determine the strength of the concrete.

In addition to primary cracks, factors affecting the strength of the concrete include the process-dependent parameters and age and hardening conditions, as well as the size and shape of the concrete body. Let us consider some of them in more detail.

Cube strength of concrete Cube specimens break under uniaxial compression by the concrete cracking in planes parallel to the compression axis. Cube-shaped concrete specimens with the cube edge size 7; 10; 15, 20 or 30 cm are suitable for compressive strength measurements. Cube specimens of $15 \times 15 \times 15$ cm in size represent the general reference. The results obtained in the testing of cube specimens require correction using an empirically determined factor β (see Table 2.2).

Also suitable for compressive strength measurements are the cylinder-shaped concrete specimens with diameters of 7; 10; 15, or 20 mm, with height twice as high as the diameter.

Prismatic strength of concrete Since reinforced concrete structures mostly have a prismatic rather than cubic shape, the prismatic strength is more common in engineering practice as well [1], [4], [6] , [8], [9]. Comparative tests of concrete prismatic vs. cubic specimens (Fig. 2.8) have shown that the prismatic strength of concrete is lower than the cubic, and diminishes with an increase in height to base edge ratio (h/a). At the value $h/a = 4$, the prismatic strength R_b stabilizes and amounts to about 0.75 of the cube strength. Table 2.3 presents average and generalized (standard) values of prismatic strength for certain grades of concrete.

Table 2.4 presents the tensile and compressive strength values of concrete applicable in designing the elements of concrete structures.

Comparison of the data presented in Tables 2.4 and 2.3 for the same types of concrete shows that the strength values applicable in engineering designing are lower than the prismatic strength values, which, in turn, are lower than cube strength values (Table 2.2).

2.4 Concrete Grades and Types: Concrete Classification [5]

The following attributes constitute the base of concrete classification:

a. Microstructure grading:

Table 2.3 Standard values of tensile and compressive strength of concrete, MPa. [8]

Deformation mode	Concrete type	Concrete class according to compressive strength (MPa)										
		B12.5	B15	B20	B25	B30	B35	B40	B45	B50	B55	B60
Uniaxial compression (prismatic strength) R_b	Heavy or fine	9.5	11	15	18.5	22	25.5	29	32	36	39.5	43
	Lightweight	9.5	11	15	18.5	22	25.5	29	–	–	–	–
Uniaxial tension R_{bt}	Heavy	1	1.15	1.4	1.6	1.8	1.95	2.1	2.2	2.3	2.4	2.5
	Fine type:											
	A	1	1.15	1.4	1.6	1.8	1.95	2.1	–	–	–	–
	B	0.85	0.95	1.15	1.35	1.5	–	–	–	–	–	–
	C	–	1.15	1.4	1.6	1.8	1.95	2.1	2.2	2.3	2.4	2.5
	Lightweight with fine aggregates:											
	Dense	1	1.15	1.4	1.6	1.8	1.95	2.1	–	–	–	–
	Porous	1	1.1	1.2	1.35	1.5	1.65	1.8	–	–	–	–

Table 2.4 Concrete strength values applicable in engineering designing, MPa. [8]

Deformation mode	Concrete type	Concrete class according to compressive strength (MPa)										
		B12.5	B15	B20	B25	B30	B35	B40	B45	B50	B55	B60
Uniaxial compression (prismatic strength) R_b	Heavy or fine	7.5	8.5	11.5	14.5	17	19.5	22	25	27.5	30	33
	Lightweight	7.5	8.5	11.5	14.5	17	19.5	22	–	–	–	–
Uniaxial tension R_{bt}	Heavy	0.66	0.75	0.9	1.05	1.2	1.3	1.4	1.45	1.55	1.6	1.65
	Fine type:											
	A	0.66	0.75	0.9	1.05	1.2	1.3	1.4	–	–	–	–
	B	0.565	0.635	0.765	0.90	1.0	–	–	–	–	–	–
	C	–	0.75	0.9	0.5	1.2	1.3	1.4	1.45	1.55	1.6	1.65
	Lightweight with fine aggregates:											
	Dense	0.66	0.75	0.9	1.05	1.2	1.3	1.4	–	–	–	–
	Porous	0.66	0.735	0.8	0.9	1.0	1.1	1.2	–	–	–	–

- dense (space between aggregate particles is completely filled by solid cement);
- macro porous or popcorn (low-sandy or high-sandy);
- high-porous (with porous aggregates and artificial porosity of solid cement)

b. Average density grading:
  - extra-heavy with density over 2500 kg/m^3;
  - heavy with density from 2200 to 2500 kg/m^3;
  - lightened with density from 1800 to 2200 kg/m^3;
  - lightweight with average density from 500 to 1800 kg/m^3;
c. Particle size grading: coarse concretes or fine concretes;
d. Aggregate grading:
  - dense aggregates;

  - porous aggregates;
  - special aggregates;
  - biologically resistant;
  - heat resistant, etc.
e. Hardening grading:
  - natural hardening;
  - heat treatment at ambient pressure;
  - high-pressure steam treatment
f. Binder grading:
  - gypsum;
  - silica;
  - acid-proof;
  - polymer-cement or polymer;
  - cement
g. General purpose in construction:
  - structural:
    - heavy concrete;
    - fine medium-density concrete;
    - lightweight concrete with dense or porous microstructure;
    - spongy concrete of high-pressure or natural hardening;
    - strained special-purpose concrete [5];
  - special concretes (heat-resistant, road, hydraulic, chemical-resistant, artificial stone, radiation-protective, heat-insulating, etc.)

Concrete quality The following classes and grades determine the main quality rating of concrete:

a. Concrete class according to compressive strength "B";
b. Concrete class according to tensile strength "B_t";
c. Frost resistance grade "F";
d. Water tightness grade "W";
e. Average density grade "D";
f. Self-stressing grade of strained concrete "S_p"

The most valuable feature of concrete is its high compressive strength, which finds wide application in concrete and reinforced concrete structures. Compressive strength is the principal parameter determining grades and classes of concretes.

The compressive strength class "B" of a concrete reflects the guaranteed compressive strength in MPa. The concrete class limits reflect the variability of measured concrete strength values with the rated coefficient of variation 13.5 %. The grade of heavy concretes expresses the ultimate compressive strength in kilogram-force per square centimeter (kgf/cm^2) measured using the reference cube specimens with edge size 15 cm after 28 days of "standard" hardening at a temperature of $20 \pm 2\,°C$ and relative humidity 80...100 % [9].

2.5   Fracture Toughness Characteristics of Concretes

Fracture toughness (or crack growth resistance) is the characteristic of material determining its resistance against breakage by crack propagation [10]. The fracture toughness characteristic is especially important for concretes since they are brittle materials and cracking is an intrinsic phenomenon for them. Many reinforced concrete structures, therefore, work in the presence of cracks. The crack opening displacement is therefore one of the primary parameters determining the permissible conditions for the use of a reinforced concrete structure, including the presence of corrosive operation environments (see Chap. 3 and [11]). In this view, the critical crack tip opening displacement δ_{IC} proposed in [12] as a criterion of fracture toughness has a particular importance for the concrete and is one of its principal physical and mechanical characteristics.

The maximal (critical) value of the stress intensity factor (SIF), K_{IC}, is another important and commonly accepted characteristic of concretes that determines the highest permissible stress state near a cleavage crack tip in a strained body. The limiting value of this factor indicates that the stress state near the crack tip in the body (concrete structure) containing the crack proportional to K_I has risen up to the highest possible value K_{IC}, and the crack begins catastrophic propagation with possible complete breakage of the structural element. K_{IC} values are subject to experimental measurement [13].

Besides the above-mentioned, fracture mechanics applies certain other parameters, such as the critical value of specific fracture energy per fresh surface formation unit (γ), critical intensity of the release rate of strain energy (G_{IC}), as well as the critical value of Cherepanov-Rice's J-integral [13]. A simple interrelation exists between all these characteristics of fracture toughness valid for conditions of plane strain and localized plasticity near the crack tip:

$$2\gamma = \sigma_0 \delta_{IC} = J_{IC} = \frac{1 - \nu^2}{E} K_{IC}^2, \qquad (2.5)$$

where σ_0 is the ultimate tensile stress (strength) of the material [12]; ν is the Poisson ratio, and E is the elastic modulus of concrete experimentally determined using respective techniques [14], [15].

Methods of determination of fracture toughness characteristics for concretes generally include [13]:

a. Methods based on experimental data on crack growth and theoretical solutions of respective mathematical crack theory problems;
b. Direct methods for determination of fracture toughness;
c. Methods based on establishing the correlating relationships between the fracture toughness and service characteristics easily measurable by standard techniques (e.g., hardness, ultimate strength, ultimate yield point, impact toughness, etc.).

Implementation of these methods requires specimens of respective shape and size with artificial cracks, as well as the necessary accuracy of crack size and critical load measurements in experiments.

The shape and size of specimens is, in many cases, dictated by the mode of material service in a structure and operational conditions.

Methods of the first group (a) are most common in practice. They are based on the following principle. Solution of a boundary problem for a cracked body of strictly defined shape and size in the mathematical crack theory yields a dependence of the form:

$$K_I = PF(l, \ E, \ v, \ H_1, \ H_2, \ H_3),\qquad(2.6)$$

where F (l, E, v, H_1, H_2, H_3) is a known function; l is crack length; $H_1, H_2,$ and H_3 are geometric parameters of the specimen; P is the applied load.

The specimen is manufactured in accordance with the mathematical conditions. The load application to this specimen under a minor loading rate allows for registering the critical load value $P = P_C$ corresponding to the beginning of crack propagation and calculating the critical value of the stress intensity factor K_I using Eq. (2.6). So obtained, the SIH value at the load $P = P_C$ is nothing other than the critical stress intensity factor K_{IC} for the chosen material.

In order that the fracture toughness value K_{IC} be representative of the material rather than the tested specimen, the stress state of the material near the crack tip must exactly correspond to the mathematical model. One condition of such correspondence consists of a specimen with thickness H_3 large enough to obey the relation [13]:

$$H_3 \geq \beta_0 \frac{K_{IC}^2}{\sigma_{0.2}^2},\qquad(2.7)$$

where coefficient β_0 is subject to experimental or theoretical determination.

Under violation of condition (2.7), the Eq. (2.6) yields a conditional critical stress intensity factor K_C that varies depending on the thickness of the specimen.

The following specimen and loading configurations are the most widely used: prismatic eccentrically tensile specimen with lateral crack (Fig. 2.6a); prismatic specimen for three-point bending (Fig. 2.6b); cylindrical specimen with outward circumferential crack for axial tension (Fig. 2.6c) or three-point bending (Fig. 2.6d); cylindrical specimen with inner crack for transverse compression along the crack plane (Fig. 2.6e).

Determination of another characteristic of fracture toughness, the specific fracture energy γ, is more convenient when using direct experiments. This method implies measurements of the work A spent for the crack length increment Δl. The loading mode must ensure a stable crack growth. In this case, the spent work (or consumed energy) is equivalent to the area under curves of loading a specimen with crack length $2 \ l$ and unloading the specimen with crack length $2 \ l + 2\Delta l$ in the experimental diagram of load P vs. crack opening u (Fig. 2.7).

The fracture toughness characteristic γ results from the formula:

$$\gamma = \frac{A}{4\Delta l\, H_3}.\qquad(2.8)$$

The advantage of such an approach is in the freedom from any mathematical solutions, which is important for engineering practice and measurement of independent fracture

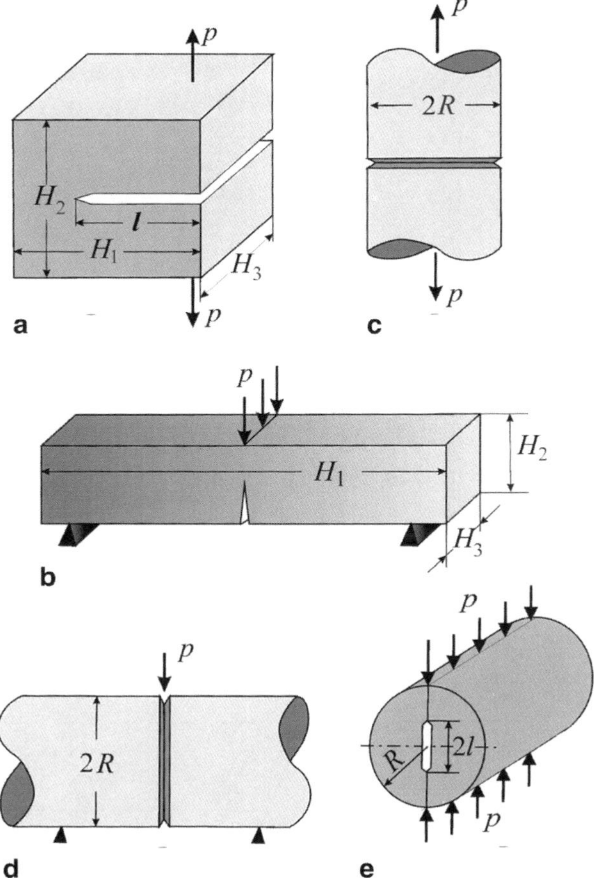

Fig. 2.6 Concrete specimen geometry for the fracture toughness experimental determination

toughness. The evaluation method of the fracture toughness characteristic, J-integral, closely resembles the determination method of γ. A direct interrelation exists between both methods within linear fracture mechanics: $2\gamma = J_{\mathrm{IC}}$.

The third group (c) of methods for measuring fracture toughness envisages measurements of standard mechanical properties and microstructure parameters with further calculation of the required value using respective correlation expressions. The analytical formula for calculation of K_{IC} derived in [13] can serve as an example of this approach:

$$K_{IC} = \sqrt{\frac{\rho \tau_T E}{(1 - v^2)\varepsilon_C}}, \tag{2.9}$$

where ρ is an intrinsic microstructure parameter; ε_c is the ultimate tensile strain; and τ_T is the ultimate shear stress of the material.

Monograph [13] contains more information about the evaluation methods for fracture toughness.

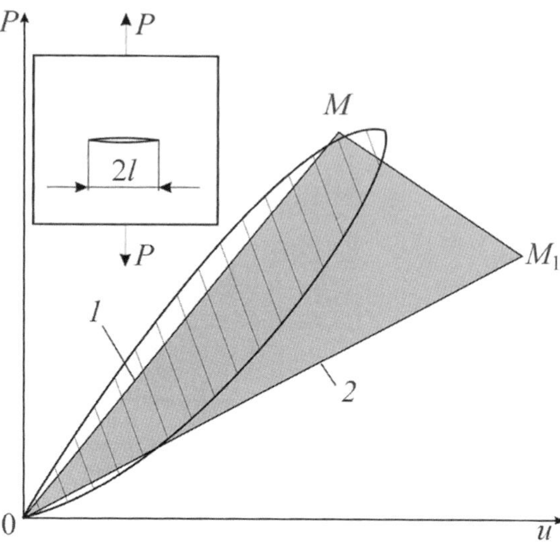

Fig. 2.7 Experimental diagram load–crack opening: curve 1 depicts loading, curve 2 unloading

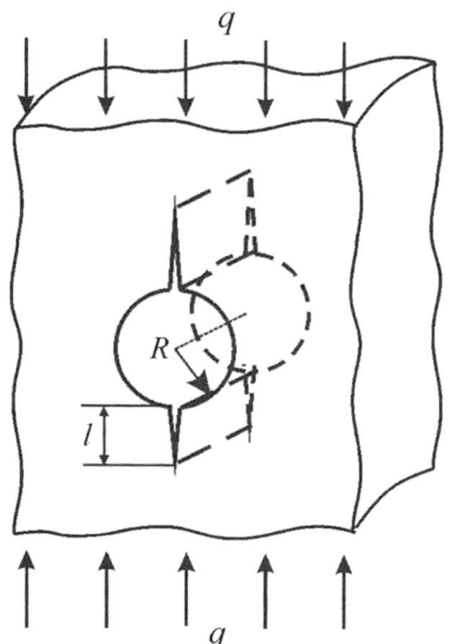

Fig. 2.8 Cracks nucleated near pore under compression

2.6   Compressive and Tensile Strength of Concrete Prisms[1]

Concrete is an inhomogeneous material, a conglomerate of cement stone and aggregates of various sizes and hardness. The compressive or tensile strength is one of the most important physical and mechanical characteristics of concretes. At present,

---

[1] The authors are grateful to N.V. Onischak for participation in the obtaining of the present results.

concrete science can formulate only individual aspects of a strength and deformation theory of concrete. The following is important in this regard.

Solids with perfect microstructure are indestructible in compression because, in their case, interactive forces between their constituents grow infinitely. A solid breaks down only when areas of tensile stress and strain arise during deformation and its constituents have the opportunity to separate one from another, with interactive (attraction) forces diminishing as the distance increases (after some critical distance). If bond stresses (interactive forces between constituents in an inhomogeneous material) begin to diminish while tensile stresses still grow, then, in some circumferences, a limiting equilibrium state arises between the constituents. That is, increments of infinitesimal tensile stress cause fractures in the given area, i.e., breakage between neighboring solid constituents. In such a way, microcracks nucleate in the above area. Further fracture development proceeds by growth and propagation of these cracks. A theory of inhomogeneous body strength based on similar concepts must derive a certain relation between tensile and compressive strengths of a concrete being the inhomogeneous material. Let us consider a simple (approximate) model of a deformed inhomogeneous body in order to obtain such a relation.

Cement stone is a key factor in determining the strength of concrete [1]. The structure of concrete as a composite material has three levels of hierarchy, namely [5], [16]: microstructure (structure of the cement stone); mesostructure (structure of cement/sand mortar), and macrostructure (structure of aggregate/mortar composite). Each level introduces its own contribution into the strength of the concrete. Analysis of these structure levels results in the following expression for the tensile strength of concrete proposed in [17]:

$$R_{bt} = A_1 R_{bt}^m = A_1 \cdot A_2 \cdot R_{bt}^c, \qquad (2.10)$$

where A_1 and A_2 are dimensionless coefficients accounting for the quality of the concrete's macrostructure and mesostructure, respectively; R_{bt}^m and R_{bt}^c are strengths of cement/sand mortar and cement stone, respectively.

Experimental measurements have shown [5] that the similar relation for the compressive strength of concrete contains the same coefficients A_1 and A_2:

$$R_b = A_1 A_2 R_b^c, \qquad (2.11)$$

where R_b^c is the compressive strength of cement stone.

It follows from Eqs. (2.10) and (2.11) that the strength of cement stone to a significant extent determines the strength of concrete. The cement stone has a capillary porous microstructure with typical pore size below $100\,\mu$m. Its tensile or compressive strength depends on porosity.

Let us accept a plate with a system of round bores with the radius R as the cement stone calculation model. Let us consider an isolated bore assuming low pore concentration so that pore interaction is negligible. Let such a plate undergo compression with uniform force q applied at large enough of a distance from the bore (Fig. 2.8).

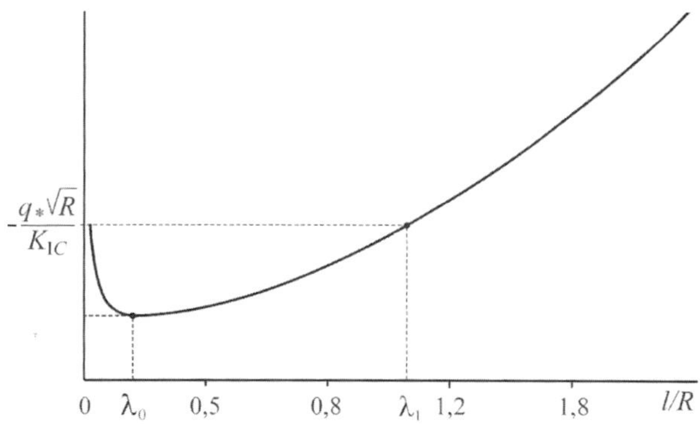

Fig. 2.9 Limiting load dependence on the crack length

Such loading creates tensile stresses at the bore circumference. A high enough load will initiate nucleation of cracks growing out of the bore (Fig. 2.8). The work [10] presents the solution of a mathematical problem for the plate with a bore and outgoing cracks. In particular, the limiting loads value for the pattern in Fig. 2.8 amounts to [16]:

$$q* = \sqrt{\frac{\pi \cdot (1+\lambda)^7}{4R[(1+\lambda)^2 - 1]}} \cdot K_{IC}; \lambda \equiv \frac{l}{R}. \tag{2.12}$$

Figure 2.9 shows the limiting load dependence on the normalized crack length l/R plotted on the base of Eq. (2.12).

The plot shows that the growth of cracks nucleated at the bore circumference under compressive load $q* = -R_b$ is initially unstable, since they propagate up to length value $l = \lambda_1 R$ requiring no loading increase. Further crack growth requires a loading intensity increase ($|q*| > -R_b$).

In this way, numerous system defects/cracks emerge in a body with bores under compression (Fig. 2.10). Since the bore spacing is large enough to neglect their interaction, interaction of cracks arranged along the compression axis is negligible as well.

So, let us suppose that our material has bores spaced at a distance $2d$ and cracks growing from the surface of each bore. The applied compressive load value q, at which the cracks merge into a single crack intercepting the whole body, corresponds to the ultimate compressive strength of the cement stone R_b^c. This material characteristic is determinable from Eq. (2.12) if we put the crack length l equal to half pore spacing d. Then,

$$R_b^c = \sqrt{\frac{\pi \cdot (1+d/R)^7}{4R[(1+d/R)^2 - 1]}} \cdot K_{IC}^c. \tag{2.13}$$

Fracture mechanics determines the tensile strength of material as:

$$R_{bt}^c = \frac{K_{IC}^c}{Y\sqrt{\pi l_0}}, \tag{2.14}$$

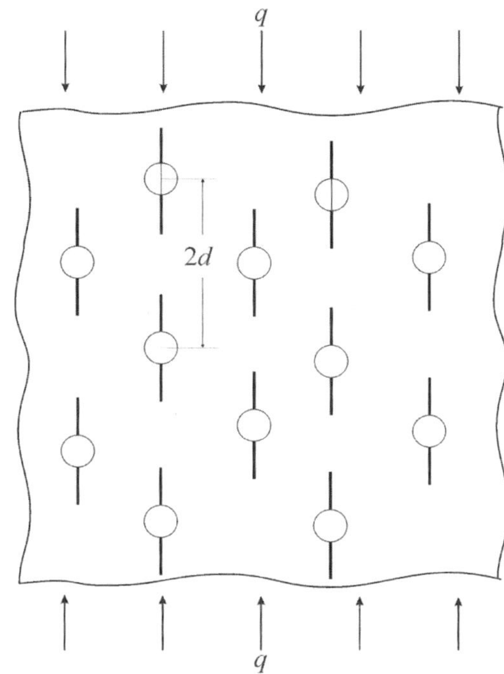

Fig. 2.10 Sketch illustration of pores with cracks in a concrete under compression

where Y is a parameter depending on crack geometry, spatial arrangement, etc.; l_0 is the half crack length assumed as being a characteristic of the material, in the given case, the cement stone. This formula allows for calculation of the crack length from experimentally measured values R_{bt}^c and K_{IC}^c,

In compliance with the above assumptions, Eqs. (2.13) and (2.14) give the following expression interconnecting the ultimate compressive and tensile strengths:

$$R_b^c = R_{bt}^c Y \sqrt{\frac{\pi l_0}{4R}} \cdot \sqrt{\frac{\pi \cdot (1 + d/R)^7}{(1 + d/R)^2 - 1}}. \tag{2.15}$$

The ultimate tensile strength of cement stone aged over 28 days can vary from 3 to 7 MPa depending on the water-cement ratio. Fracture toughness K_{IC} of the same cement stone varies within the range of $0.11\ldots 0.4\,\mathrm{MPa\cdot m^{1/2}}$. The value of coefficient Y for linear cracks in the plane problem is close to 1.0. Under such conditions, Eq. (2.14) yields the range of l_0 variation from 0.2 to 0.4 mm. Let us put a mean pore size equal to 0.1 mm. Total porosity is about 20 % of that corresponding to mean pore spacing $d \approx R$. Substitution of the above values into the relationship (1.15) permits for calculation that $R_b^c \approx (14\ldots 20)\,R_{bt}^c$ in good agreement with experimental results.

The combination of Eqs. (2.10), (2.11), and (2.15) yields a similar relation between compressive strength (R_b) and tensile strength (R_{bt}) for concretes:

$$R_b = R_{bt} \cdot \frac{\pi Y}{2} \sqrt{\frac{l_0 \cdot (1 + d/R)^7}{R[(1 + d/R)^2 - 1]}}. \tag{2.16}$$

Table 2.5 Classification and mechanical properties of concrete reinforcement. [8]

Reinforcement designation and grade	Steel grade	Cross-section diameter, mm	Yield point, MPa	Ultimate strength, MPa	Tensile strain, %
1	2	3	4	5	6
Round hot-rolled reinforcing bar, grade A-1	St3. BSt3	6…40	230	380	25
Periodic profile, grades:					
A-I I	BSt5	10…40			
	10GT	10…32	300	500	19
	18G2S	40…80			
A-I I I	25G2S	6…40			
	35GS	6…40	400	600	14
	32G2R	6…22			
A-I V	20CrG2Z	10…22			
	80S	10…18	600	900	8
A-V	23Cr2G2T	10…22	800	1050	7
A-VI	20Cr2G2SR	10…22	1000	1200	6
Heat refined reinforcing bar, grades:					
At-I I I S	BSt5SP	10…38	400	600	–
At-I VS	25G2S	10…28	600	900	8
At-V	20GS	10…22	800	1050	7
At-VI	20GS	10…22	1000	1200	6
Ordinary reinforcing wire with periodic profile, Grade Vr-I	–	3…5	–	550…525	–
High-strength reinforcing wire:					
Smooth, grade V-II	–	3…8	–	1900…1400	4…6
Periodic profile, grade Vr-II	–	3…8	–	1800…1300	4…6
Reinforcing rope:					
Grade K -7	–	6…15	–	1850…1650	–
Grade K -19	–	14	–	1800	–

2.7 Classification and Mechanical Properties of Concrete Reinforcement

The purpose of concrete structure reinforcement consists mainly in resistance to tensile forces arising in its elements. The necessary reinforcement amount is subject to estimation based on the operational loading of the structural elements.

Reinforcement classification is comprised of four classes:

a. Steel reinforcement for reinforced concrete structures, including hot rolled rod and cold-drawn wire or rod (with any diameter) reinforcement;
b. Hot rolled reinforcement, including heat refined rods or rods hardened by cold drawing, dragging, etc.;

Table 2.6 Standard and engineering design values of strength and elastic modulus for reinforcing bars. [8]

| Reinforcement grade | Standard strength R_{sn}, MPa | Strength values for engineering design, MPa | | Elastic modulus E_s, MPa |
| | | Tensile | | Compressive R_{sc} | |
		Either (a) longitudinal or (b) transverse strength per skew section under bending moment R_s	Transverse strength per skew section under transverse force R_{sw}		
A-1	235	225	175	225	210,000
A-1 1	295	280	225	280	210,000
A-1 1 1, 6…8 mm dia	390	355	285	355	200,000
A-1 1 1 and At-1 1 1, 10…40 mm dia.	390	365	290	365	200,000
A-1V and At-1VS	590	510	405	390	190,000
A-V and At-V	785	680	545	390	190,000
A-V1 and At-V1	980	815	650	390	190,000

c. According to surface relief, reinforcement may include rods of periodic profile or plain wire; periodically, ribbed profile of rods as well as ledges and dents on a wire surface essentially enhance the adhesion with the concrete;

d. According to method of use in concrete structures, reinforcement may include pre-stressed (pre-stretched) or unstressed reinforcement

The hot-rolled reinforcing rods have six grades depending on the main mechanical characteristics (Table 2.5) [8] conventionally designated as A-I, A-ĐⳆĐⳆ, A-III, A-IV, A-V, and A-VI.

Heat refined reinforcing bars have an additional symbol "t" in their designation, namely At-1 1 1, At-1V, At-V, and At-V1. The reinforcing bars intended for placement by cage welding have an additional letter "S" in their designation after the Roman numeral. If the reinforcement has an enhanced corrosion resistance, its designation has the letter "K". Cold drawing hardened reinforcement is marked with an additional letter "V" in the grade name.

Each reinforcement grade covers steels with different chemical composition but similar values of yield point (Table 2.5). Table 2.6 presents standard and engineering design values of the strength and elastic modulus for reinforcing bars.

Table 2.7 shows the same characteristics for the reinforcing wires.

Table 2.7 Standard and engineering design values of strength and elastic modulus for reinforcing wires and wire ropes. [8]

Grade	Diameter, mm	Standard strength R_{sn}	Strength values for engineering design, MPa		Compressive R_{sc}	Elastic modulus E_s, MPa
			Tensile			
			Either (a) longitudinal or (b) transverse strength per skew section under bending moment R_s	Transverse strength per skew section under transverse force R_{sw}		
Vr-	3	410	375	270	375	
	4	405	370	265	370	
	5	395	360	260	360	170,000
V-	3	1490	1240	990	390	200,000
	4	1410	1180	940	For all grades in bonding with concrete	
	5	1330	1100	890		
	6	1250	1050	835		
	7	1180	980	785		
	8	1100	915	730		
B-	3	1460	1200	970	–	200,000
	4	1370	1140	910		
	5	1250	1050	830		
	6	1180	980	785		
	7	1100	915	735		
	8	1020	850	675		
K -7	6	1450	1200	970	–	180,000
	9	1370	1140	910		
	12	1330	1100	890		
	15	1290	1080	865		
K -19	14	1410	1180	940	–	180,000

The above-presented classification and properties of reinforcing steels correspond to the Standard of Ukraine DSTU 3760-98 issued in the year 1998 [18] and supplemented in the year 2006 [19]. These new Ukrainian standards introduced the classification adapted to international standards ISO 6934, ISO 6935, DIN 488, and ENU 10080 in respect to size, chemical composition, mechanical characteristics, and testing methods.

The new standard divides the reinforcing bars (A) into grades according to the physical yield point expressed in MPa (N/mm^2).

Finally, according to service characteristics, reinforcement can be as follows:

- Welding (letter "S");
- Stress corrosion cracking resistant ("K");
- Unweldable (without the letter "S");
- Corrosion nonresistant (without the letter "K")

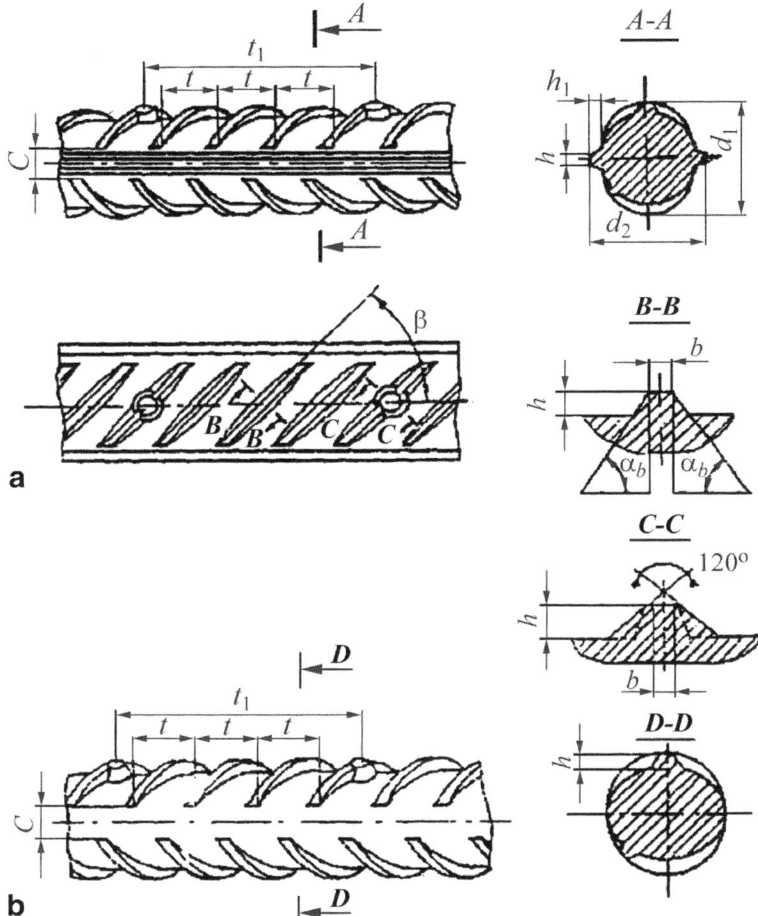

Fig. 2.11 Geometry of periodic hardened profile for rolled reinforcing bars

Ukrainian manufacturers produce ten different grades of reinforcing bars, namely: A240C with plain profile; A300C, A400C, A500C, A600, A600C, A600K, A800, A800K. and A1000 with periodic hardened profile (Fig. 2.11). Table 2.8 exhibits the technical characteristics of the reinforcing bars.

In the past, yield point was the only criterion of a steel grade selection for producing the reinforcement of a specified grade (see Table 2.5). Now, in compliance with Ukrainian standard DSTU 3760-98, the reinforcement production procedures rely on chemical elemental weight composition specified for each reinforcement grade (Table 2.9).

Weight percentage of other alloying elements (chromium, copper, nickel) for welding reinforcement should be less than 0.3 % for each. Chromium weight content 0.90 % or less is admissible for high-strength reinforcing bar grade A800K or A1000.

Table 2.8 Standard values of diameter, cross-section area, and running meter mass for reinforcing bars. [18]

Standard rod diameter d_n, mm	Standard cross-section area, mm^2	Running meter mass of reinforcing bars:	
		Design value, kg	Tolerance, %
1	2	3	4
5.5	32.8	0.187	
6.0	28.3	0.222	± 8.0
8.0	50.3	0.395	
10.0	78.5	0.617	
12.0	113.0	0.888	± 5.0
14.0	154.0	1.210	
16.0	201.0	1.580	
18.0	254.0	2.000	
20.0	314.0	2.470	
22.0	380.0	2.980	
25.0	491.0	3.850	± 4.5
28.0	616.0	4.830	
32.0	804.0	6.310	
35.0	1018.0	7.990	
40.0	1255.0	9.860	

Running meter mass of bars in kg is calculated based on nominal (standard) diameter and steel density 7.85 MT/m^3

Table 2.9 Chemical composition of reinforcing steels. [18]

Reinforcing bar grade	Highest permissible content of chemical elements, wt. %						
	Carbon	Silicon	Manganese	Phosphorus	Sulfur	Nitrogen	Arsenic
A240S	0.22	–	–	0.045	0.045	0.012	0.08
A300S	0.22	–	–	0.045	0.045	0.012	0.08
A400S	0.22	–	–	0.045	0.045	0.012	0.08
A500S	0.22	–	–	0.045	0.045	0.012	0.08
A600; A600S; A600K	0.28	1.00	1.6	0.045	0.045	0.012	0.08
A800; A800K	0.32	2.40	2.3	0.040	0.040	0.012	0.08
A1000	0.32	2.40	2.3	0.040	0.040	0.012	0.08

The most recent Ukrainian standard DSTU 3760:2006 (Appendix A) [19] defines specific steel grades appropriate for producing reinforcing bars of certain grades (see Table 2.10).

Note that the standard DSTU 3760:2006 (Appendix D) [19] establishes the following corrosion cracking requirements for the corrosion resistant grades of reinforcing bars. The time for a corrosion cracking rupture of a corrosion-resistant reinforcing bar specimen in the nitrate solution consisting of calcium nitrate, ammonium nitrate and water in weight proportion 600:50:350 must be at least 100 h at a temperature of 98...100 °C and at applied stress $0.9\sigma_{0.2}$ (as calculated from Table 2.11).

Table 2.10 Steel grades recommended for producing reinforcing bars. [19]

Reinforcing bar grade	Steel grades acc. Ukrainian standards DSTU 2651:2005 (GOST 380-2005, GOST 5781-82, GOST 10884-94)	Bar forming method	Bar diameter, mm
A240S	St3sp, St3 ps, St3kp	Hot rolled	5.5–40
A400S	St3sp, St3 ps, St3Gps, St5sp, St5 ps, 25G2S, 35GS	Thermomechanically hardened hot rolled	6–40
A500S	St3sp, St3 ps, St3Gps, St5sp, St3Gps, 25G2S	Thermomechanically hardened	6–16, 18–22, 25–40
A600	20GS	Thermomechanically hardened	10–32
A600S	25G2S, 35GS		
A600K	10GS2. 08G2S		
A800	20GS, 20GS2, 08G2S, 10GS2	6–40	
A800K	35GS		
A800SK	20CrGS2		
A1000	25G2S, 20CrGS2	6–40	

Table 2.11 Mechanical characteristics of reinforcing bars. [18], [19]

Reinforcing bar grade	Temperature of electrical heating, °C	Ultimate tensile strength σ_B, MPa	Ultimate yield point $\sigma_{0.2}$, MPa	Ultimate tensile fracture strain δ_s, %	Ultimate uniform fracture strain δ_p, %	Total strain at maximum loading δ_{max}, %	Initial elastic modulus $E \times 10^{-4}$, MPa	Bending angle during bending tests, deg.	Boring bar diameter (in relation to nominal bar diameter d_n)
				Equal or higher than					
A240S	–	370	240	25	–	–	21	180	$0.5\,d_n$
A300S	–	490	290	19	–	2.5	21	180	$3\,d_n$
A400S	–	500	400	16	–	2.5	20	90	$3\,d_n$
A500S	–	600	500	14	–	2.5	19	90	$3\,d_n$
A600									
A600S	400	800	600	12	4	2.5	19	45	$5\,d_n$
A600K									
A800									
A800K	400	1000	800	8	2	3.5	19	45	$5\,d_n$
A1000	450	1250	1000	7	2	3.5	19	45	$5\,d_n$

However, maximum permissible fracture toughness values for both corrosion static and corrosion cyclic cracking remain unrestricted, although these characteristics are very important, especially for structures in long-term operation requiring rehabilitation [20].

Table 2.11 demonstrates the mechanical properties and elastic modulus values for various grades of reinforcing bars.

References

1. Berg OY (1961) Fizicheskiye osnovy prochnosti betona i zhelezobetona (Physical fundamentals of concrete and reinforced concrete strength). GSI, Moscow
2. Vinogradov BN (1979) Vliyaniye zapolniteley na svoystva betona (The effect of aggregates on concrete properties). Stroyizdat, Moscow
3. Bazhenov YM (1978) Tehnologiya betona (The concrete technology). Vyssh.shkola, Moscow
4. Ahverdov IN (1981) Osnovy tehnologii betona (Basics of concrete physics). Stroyizdat, Moscow
5. Grushko IM, Illyin AG, Chikhladze ED (1986) Povysheniye prochnosti i vynoslivosti betona (Advance in concrete strength and durability). Vyssh.shkola, Moscow
6. Gotz VI (2003) Betony i budivel'ni rozchiny (Concretes and building mortars). KNUBA, Kyiv
7. Shishkin AA (2001) Fizicheskaya khimiya kontaktnoy zony betonov (Physical chemistry of contact zone in concretes). Mineral, Krivoy Rog
8. Baikov VN, Sigalov EE (1984) Zhelezobetonnyye konstruktzii (Reinforced concrete structures). Stroyizdat, Moscow
9. Petzold TM, Tur VV (2003) Zhelezobetonnye konstruktzii. Osnovy teorii rascheta i konstruirovaniya (Reinforced concrete structures: basics of design and development). BGTU, Brest
10. Panasyuk VV (1991) Mehanika kvazikhrupkogo razrusheniya materialov (Mechanics of quasi-brittle fracture of materials). Nauk. dumka, Kyiv
11. Moskvin VM, Alexeev SA, Verbitskiy GP, Novgorodskiy VI (1971) Treshchiny v betone i korroziya armatury (Cracks in concrete and corrosion of reinforcement). Stroyizdat, Moscow
12. Panasyuk VV (1968) Predel'noye ravnovesiye khrupkikh tel s treshchinami (The limiting equilibrium of brittle bodies with cracks). Nauk. dumka, Kyiv
13. Panasyuk VV, Andreikiv AE, Kovchik SE (1977) Metody otsenki treshchinostoykosti konstruktsionnykh materialov (Methods of fracture toughness estimation in structural materials). Nauk. dumka, Kyiv
14. Betony. Metody opredeleniya prizmennoy prochnosti, modulya uprugosti i koefficienta Puassona (1980) Concretes: methods for determining the prismatic strength, elastic modulus, and Poisson ratio. Gos. kom. po delam stroitelstva, Moscow
15. Zaitsev YV (1982) Modelirovaniye deformatsiy i prochnosti betona metodami mekhaniki razrusheniya (The concrete deformations and strength simulation using fracture mechanics methods). Stroyizdat, Moscow
16. Standard of Ukraine DSTU 3760-98. Prokat armaturnyi dlya zhelezobetonnykh konstruktsiy (1998) Reinforcing bars for concrete structures. Gosstandart Ukrainy, Kyiv
17. Grushko IM, Illyin AG, Rashevskiy ST (1965) Prochnost' betonov na rastyazheniye (Tensile strength of concretes). Vyssh.shkola, Kharkiv
18. USSR Standard GOST 29167-91. Betony. Metody opredeleniya treshchinostoykosti (vyazkosti razrusheniya) pri staticheskom nagruzhenii (1994) Concretes: methods for determining fracture toughness under static loading. Gosstroyizdat, Moscow
19. Standard of Ukraine DSTU 3760:2006. Prokat armaturnyi dlya zalizobetonnykh konstruktsiy. Zahal'ni tekhnichni umovy (2006) Reinforcing bars for concrete structures: general technical conditions. Derzhspozhivstandart Ukrainy, Kyiv
20. Romaniv ON, Nikiforchin GN (1986) Mekhanika razrusheniya konstruktsionnykh splavov (Fracture mechanics of structural alloys). Metallurgiya, Moscow

Chapter 3
Predominant Damages and Injuries in Reinforced Concrete Structures Arising During Use

Abstract The subject of this chapter is the effect of main corrosive air and/or water environment factors on strength of structural concrete and reinforced concrete during long-term operation. Specific destructive action of static and/or cyclic loads and induced complex deformations on processes of concrete damage and fracture is considered. Included case studies characterize decrease in concrete strength and load-carrying capability due to intense corrosion processes occurring on its surface and/or bulk. The specific effect of aggressive gas environments causing discontinuity of concrete's heterogeneous structure and nucleation of defects such as cracks, laminations, cavities or others is characterized. Outline and characterization of principal biological and microbiological factors affecting the concrete injury in practice are given. Basic kinds of stress corrosion degradation causing thinning and load-bearing capacity diminishment of steel reinforcing wires in concrete are covered. The authors present analysis of combined effects of static/cyclic mechanical loads and corrosive air and/or water environment on injury and fracturing of concrete structures and constructions over long-term operation.

3.1 Specification of the Effect of Environmental Factors on the Strength of Concrete

Concrete and reinforced concrete structures and buildings over long-term use experience the cumulative influence of many physical, chemical, and other factors of assorted nature and/or intensity that reduce their strength and service life [1], [2]. Most often, damage, injury, and fracture in buildings and structures[1] originate from a corrosive environment resulting in corrosion of the concrete and reinforcement. Prolonged superimposed action of mechanical stresses in turn generates defects in the materials and causes stress corrosion cracking. Therefore, the strength and service life of concrete and reinforced concrete structures is the result of the combined effects of mechanical loads and environment accumulating over time. The main destructive effects of external factors may have a mechanical, chemical, thermal, or biological nature [3], [4]. The mechanisms of these effects involve respective natural phenomena such as local material fracture (damage) caused by stress concentration near

---

[1] The term 'structures' hereinafter implies concrete or reinforced concrete structures, except when otherwise specified.

V. V. Panasyuk et al., *Injection Technologies for the Repair of Damaged Concrete Structures,* DOI 10.1007/978-94-007-7908-2_3,
© Springer Science+Business Media Dordrecht 2014

the structural singularities; corrosive injury caused by chemical interaction of the environment with the constituents of the concrete and/or reinforcement; or thermal failure due to temperature gradients as well as biological injury by living organisms (e.g., bacteria).

Biological corrosion, in particular, results from the harmful vital activity of bacteria and requires specific protection and repair measures for building structures [5], [6].

In the past, many famous scientists and engineers have put great effort into solving the above problems [6], [7]. Unfortunately, the versatile nature of damaging processes, especially the corrosive processes, precludes establishing a single effective prevention method. Mechanisms of damage formation in strained materials are subject, hence, to further study, as well as the main factors affecting these processes and possible preventive measures.

For example, V.M. Moskvin defines the following three predominant mechanisms out of a great number of assorted corrosive processes [1], [2].

The first mechanism describes concrete corrosion by moderately hard aqueous solutions. The flowing water dissolves and subsequently removes certain constituents of the cement stone.

The second mechanism includes exchange reaction between other constituents of the cement stone and impurities present in water.

Finally, the third mechanism of corrosion assumes failure of the cement stone due to precipitation and crystallization of poorly soluble salts and water residues in voids or capillaries of the cement stone, as well as cavities within the concrete with subsequent freezing as the temperature lowers. These processes cause significant mechanical stresses resulting in breakage of the concrete constituents. This mechanism significantly contributes to mechanical factors and causes cracking of concretes. Water volume increase during crystallization (freezing) in the concrete micro voids is by itself the primary source of high stresses that are responsible for local concrete failure and the formation of overstressed zones.

3.2 Mechanical Loading Effect on Concrete Damage and Fracture

Externally applied loads, especially pulsing, cycling, and statically prolonged, are a very important factor affecting the formation of damage in concrete or reinforced concrete structures. Many concrete structures such as dams, bridges, framed trestles, road pavements, etc., operate under the same commonly known conditions [8], [9].

This section provides a separate consideration of each loading mode influence on concrete properties, damaging, and fracture resistance.

Long-term Static Compressive or Tensile Loads

The mechanical properties and elastic modulus of concretes significantly depend on the age of the concrete, especially during the first six months of its hardening.

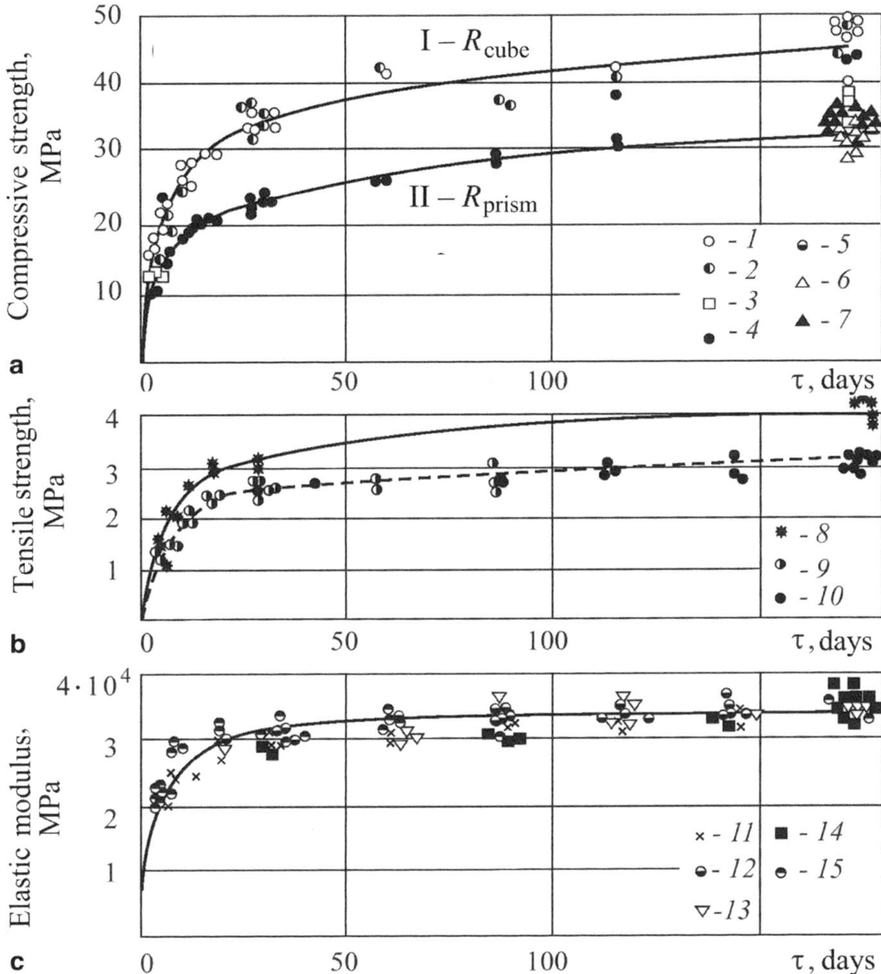

Fig. 3.1 Concrete age dependence of ultimate compressive and tensile strengths as well as elastic modulus for specimens of different shape and size [9]: cube $7 \times 7 \times 7$ cm (*1*); cube $10 \times 10 \times 10$ cm (*2*); cube $20 \times 20 \times 20$ cm (*3*); prisms $7 \times 7 \times 60$ cm (*4, 6*); prisms $10 \times 10 \times 100$ cm (*5, 7*); waisted specimens $5 \times 5 \times 28$ cm (*8*); waisted specimens $10 \times 10 \times 100$ cm (*9*); prisms $10 \times 10 \times 100$ cm (*10, 14*) under compression and tension in various sequences, and prisms under step cycling loading (*15*)

Book [9] presents results of experimental studies for four series of concrete specimen made from the same brand of cement but with different cement-sand-crushed rock (c/s/r) and water-cement (w/c) ratios. Figure 3.1 shows the experimental time dependence of compressive and tensile strengths, as well as the elastic modulus for specimens with the following solid concrete composition: $c/s/r = 1{:}2.63{:}3.59$ and $w/c = 0.622$. One can easily see from the figure that the compressive strength of the cube specimens $10 \times 10 \times 10$ cm in size was 35.1 MPa after 28 days of exposure but 44.6 MPa after 175 days.

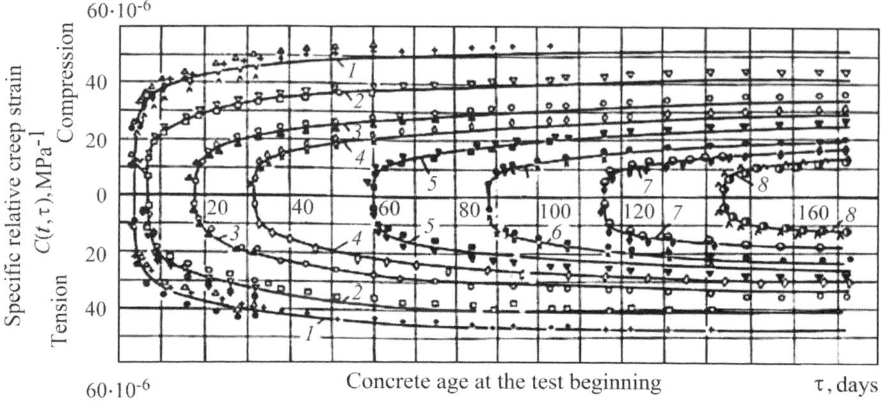

Fig. 3.2 Dependence of specific relative creep strains for identical concrete specimens on loading time τ elapsed after hardening time *t* equal to 4, 7, 18, 32, 60, 88, 116, or 144 days (respective *curves*)

Further exposure leads to slower and slower strength growth asymptotically approaching the respective values for old concrete.

The elastic modulus growth (Fig. 3.1c) is similarly fastest during the initial hardening period 2.1c) with a duration of 18 days. After this hardening time, Young modulus values amount to about 0.9 of the limiting values for old concrete at compression or 0.95 of the limiting values at tension.

According to [11], the best fitting function for Young modulus dependence on the hardening time of the concrete is:

$$E(\tau) = E_0(1 - e^{-\alpha\tau}), \tag{3.1}$$

where E_0 is the limiting value of the Young modulus for old concrete; α, β are empirically determined constants. In particular, insertion of the experimental data presented in Fig. 3.1 gives the following values of parameters: $E_0 = 3.3 \cdot 10^{6\text{MPa}}$; β = 0.53, and α = 0.08 day⁻¹.

Long-term loads also induce concrete creep processes. The concrete creep theory pays most attention to so-called 'ordinary creep', i.e., concrete deformation under load over prolonged periods. If comparing concrete creep under compression or tension, the tensile creep strain will be 3–5 times higher than the compressive strain at the same stress level. If comparing creep strains at the same percentage of ultimate strength, conversely, the compressive strain will be considerably higher than the tensile strain. However, some opposite data exist [9] stating that, at the same percentage of ultimate strength, the creep strain values can be equal or even less under compression.

The work [9] also proposes an effective creep characteristic, which has the form of specific relative strain: $C(t,\tau) = \varepsilon/\sigma = \Delta l/l_\sigma[\text{MPa}^{-1}]$. Figure 3.2 illustrates the typical time dependences of specific lative strain for concrete specimens under constant stress.

Compressive stresses here were maintained within $(0.32\dots0.39)R_b$ whereas tensile stresses were within $(0.62\dots0.73)R_{bt}$. Figure 3.2 shows that the relative creep strains under tension are approximately $5\dots20\,\%$ higher than the strains under compression. Prolongation of hardening time from 4 to 144 days till applying creep load results in specific relative strain decrease by approximately 3–5 times.

Concretes usually demonstrate an elastic aftereffect that show recovery of initial specimen sizes after unloading. More specifically, the elastic aftereffect strain is about $60\,\%$ of the creep strain, as shown by testing identical specimens of mature aged concrete. Consequently, unloaded specimens after either long-term tensile or long-term compressive loading would reveal only partial reversibility of creep strains, tensile loading providing higher recovery. In conclusion, concrete is not a purely elastic material, with all resulting consequences.

Many authors, in particular O. Berg [10], believe that concrete creep strain is due to nucleation and growth of microcracks under applied loads. The above-mentioned author [10] as well as Arutyunyan [11] have shown that, beyond a certain stress level, irreversible cracks appear in the concrete with the crack edges remaining open after unloading. During further loading, such cracks cause concrete fracture.

It follows from the above results that only injection materials with a high enough permeability for filling the opened cracks are appropriate for renewal of structures of long-term operation.

Cycling Loads

Today, concretes and reinforced concretes find wide applications in structures operating under cyclic loading, such as bridges, roads, railway sleepers, machinery foundations, crane ways, machine-building structures, etc. Numerous studies [12] have established that high-cycled repeated loads break concrete at stresses considerably lower than its ultimate strength under static loading.

The relative fatigue strength $k = R_{BT}/R_b$, where R_{BT} is the fatigue strength, is within $0.47 < k < 0.60$ at the stress ratio $\sigma_{min}/\sigma_{max} = 0.8$ as indicated by many authors [9], [12]–[14].

When considering fatigue or endurance strength, one should remember that concrete is a specific material. Its physical and mechanical characteristics strongly depend on rates of loading and ageing that degrade the initial state of structure. Figure 3.3 presents the rated dependencies of the relative fatigue strength and Young modulus on the loading stress (σ_{max}) cycling frequency (ω) for concrete specimens.

One can see that the loading (strain) rate effect is most pronounced at the initial stage of cycling. The fatigue strength increases to some extent with a rise in the loading rate (Fig. 3.4). The processes of cyclic deformation of concrete structures run most intensively in zones of stress concentration, namely, near structural defects and constituents with differing physical and/or mechanical properties, in other words, interfaces between an aggregate and the cement stone. Just phase boundaries are those places where the material loosening arises, as well as its fatigue damaging, crack nucleation, etc.

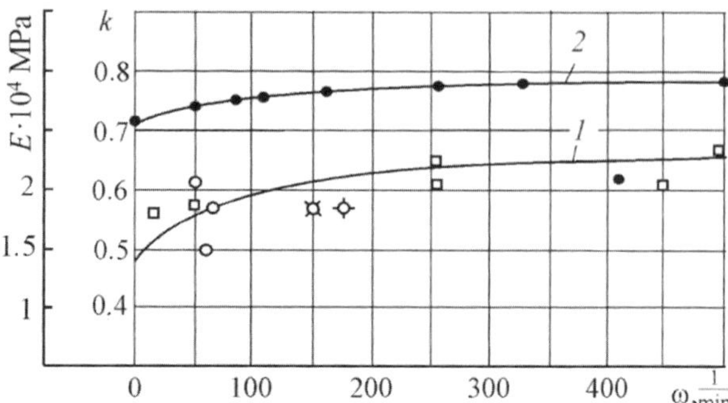

Fig. 3.3 The effect of loading frequency (ω) on relative fatigue strength (curve *1*) and Young modulus (curve *2*) for the concrete [12]

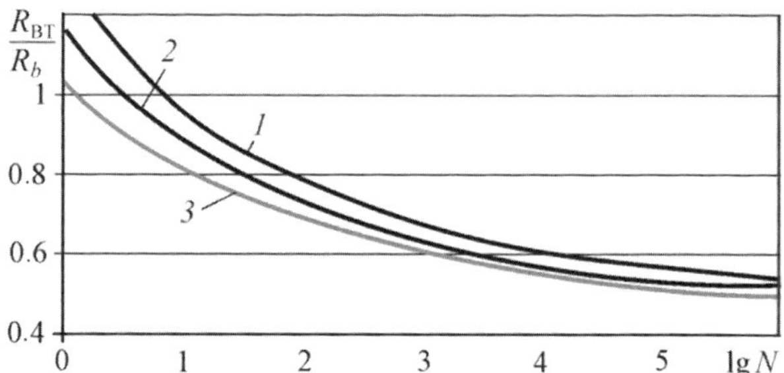

Fig. 3.4 Family of curves characterizing the fatigue resistance of concrete depending on cycling frequency: 50 Hz (*1*); 5 Hz (*2*); 0.5 Hz (*3*) [12]

Many authors [9], [10], [12], [13], [15] believe that the material plasticity in the overstressed areas plays an essential role in microcrack nucleation. Plastic shears and creep lead to stress redistribution, reducing stress concentration. For this reason, concretes with enhanced plasticity have a higher cyclic endurance under the same additional conditions of loading.

Berg [10] argues that concrete microcracking arising in the first loading cycle causes loosening of the initial material, which exerts a decisive influence on further damaging and final strength.

Thus, cyclic loading is the main mechanism of concrete damage and impairment, which results in nucleation and slow growth of cleavage microcracks (mode I). These processes occur in the places of local stress concentration formed due to concrete inhomogeneity.

3.3   Concrete Corrosive Softening and Injury

Corrosive Injury of the First Kind

Calcium hydroxide is the most soluble component of the Portland cement-based stone. Therefore, the corrosive process of the first kind concerns calcium hydroxide dissolution, or so-called lime leaching. Concrete resistance to this kind of corrosion depends on the hydrolytic stability of cement stone minerals, which, in turn, depends on the concrete permeability (density). Studies of cement stone's chemical behavior and chemical stability in various conditions demonstrated the danger of leaching corrosion as early as the beginning of the twentieth century. In particular, certain authors [16] have stated that all concrete structures made from Portland cement are liable to lime leaching and will inevitably crumble after a while. In the absence of necessary protective facilities, all structures made from Portland cement under long-term operation in a water environment are doomed to failure. Such long-term operation under permanent influence of water leads to lime leaching. White deposits on the concrete surface are exterior expressions of leaching. White spots on concrete structures or buildings are leached lime hydrate deposits and indicate that these structures are based on the Portland cement. For example, such deposits indicate lime leaching in gaps between the granite tiles on the concrete column that stands as a monument to the soldier-liberators in Lviv, Ukraine (see Fig. 3.5 of Appendix).

Many concrete injuries of the first kind had appeared on the dam body of the Tashlytska hydroelectric pumped storage power station, Ukraine (Fig. 3.6a of Appendix) after filling the water storage basin with water (Fig. 3.6b of Appendix). Over 20 years passed from the beginning of the dam's construction to the dam's start-up. In the course of soil settlement, the gaps or cracks appeared between concrete blocks in the dam's body, openings that allowed for water leakage from the upper water storage. Dissolved lime formed white stains while streaming down the exterior wall of the dam (Figs. 3.6c, d of Appendix). Such gaps or cracks are subject to repair by injection technologies, in particular, sealing methods.

Tunnels (Fig. 3.7 of Appendix) and bridges (Figs. 3.8 and 3.9 of Appendix) break under the combined action of the first kind of corrosion and mechanical stress.

In general, running water that seeps through concrete essentially accelerates the softening of the concrete due to the first kind of corrosion. The corrosion rate is directly proportional to the water seepage rate within certain limits. Figure 3.10 illustrates the concrete leaching with clear water and shows three typical stages of the process: 0–1; 1–2, and 2–3 inherent to different leaching conditions. The directly proportional dependence between the amount of dissolved lime and the volume of seeped water exists in stage 0–1. Here, the volume of water seeped through the concrete surface unit per time unit determines the rate of concrete corrosion and, hence, the extent of damage to the structure.

Stage 1–2 corresponds to another filtration condition where the dissolved lime concentration grows slower with an increase in the filtration rate because the process is limited by the lime diffusion from gel capillary depth into the water filtration flow.

Finally, stage 2–3 represents the amount of dissolved lime roughly independent of the volume of seeped water. In this stage, the dissolution rate is determined only by $Ca(OH)_2$ diffusion rate in the cement stone. Such a process occurs when great volumes of water wash the concrete surface. The conditions inherent to the first stage are encountered quite rarely and only in large structures during the initial period of filtration. Here, the lime leaching becomes apparent as white sediments. Later, the filtration dramatically slows due to concrete self-sealing and a decrease in the amounts of dissolved lime.

Fig. 3.5 a Concrete column of monument to soldier-liberators at Stryis'ka Str. in Lviv, Ukraine. **b** Lime deposits on granite tiles (roughly 40 years old)

Fig. 3.6 a, **b** Overall view of the dam body of the Tashlytska hydroelectric pumped storage power station, Ukraine. **c**, **d** White stains of lime leached from concrete

Fig. 3.7 a Damages in road tunnel. **b** Appearance of open stress corrosion cracks in the tunnel

Figure 3.11 presents the leaching degree as a function of the volume of seeped water.

The plot summarizes the results of [2]: Ivanov for cement stone (*1*) and concrete (*2*); Hellstrom for cement stone (*3*); Rutgers for cement stone at $K_F = 10^{-4} \ldots 1.5{\cdot}10^{-5}$ cm/s (*4*) or $3.5 \ldots 6 \cdot 10^{-5}$ cm/s (*5*); Li for cement mortar (*6*); calculation for the solution concentration 1.36 g/l (*8*), and calculation for catastrophic strength loss (*9*).

Fig. 3.8 Appearance of bridge pillars damaged by cleavage cracks caused by cyclic compression (**a**), cleavage/shear cracks (**b**), and fissures in plaster

All experiments with concretes had revealed a slowing of the leaching processes after removing the 15 g of lime per 1 kg of initial concrete, which indicates diffusion as being the rate-controlling mechanism.

Fig. 3.9 Three-span arch bridge (**a**) damaged by cracks (**b**, **c**) due to concrete weakening and cyclic loads

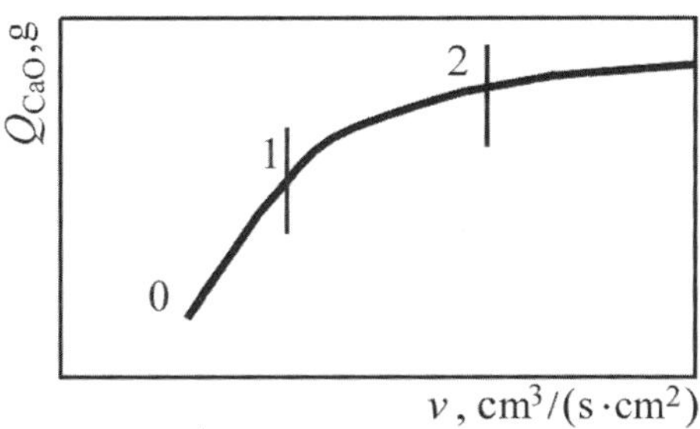

Fig. 3.10 The dependence of lime amount removed from concrete from the water filtration rate [2]

Calculations of lime leaching and permissible water filtration coefficient (K_F) or concrete strength retention time during either water seepage through the concrete or water pouring onto the concrete surface may serve as quantitative measures of the first kind of corrosion and the base for predictions of a concrete's service life. Simplification of such calculations is possible if the data on lime removal rate q (see Fig. 3.11), filtration coefficient K_F, and the permissible degree of cement stone leaching based on the permissible strength R_b variation are known for a given specific situation. The known R_b dependence on the lime leaching degree q_{lime} enabled us to establish the permissible concrete depletion with CaO, which should be less than 10 % since the strength of the concrete abruptly decreases and the state of the cement stone becomes unstable beyond this limit.

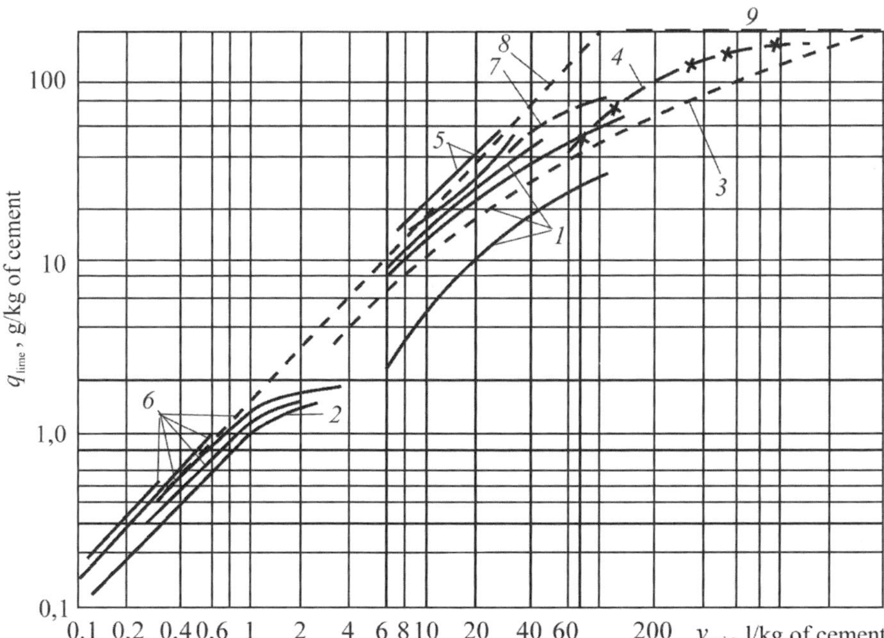

Fig. 3.11 Lime share leached from cement stone or concrete as a function of seeped water volume

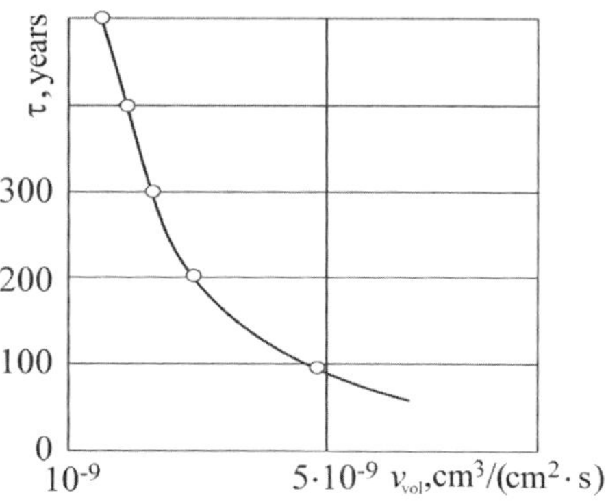

Fig. 3.12 Dependence of concrete service life (resource) on the water filtration rate [2]

In the above approximation, one can write [2]:

$$\tau = \frac{q_{\text{lime}}}{V_{\text{vol}}} \cdot C_{\text{lime}}, \tag{3.2}$$

where q_{lime} is the amount of lime weight that can be removed from concrete without mechanical deterioration; C_{lime} is mean lime concentration in water over service life; τ is the summary water action time until reaching the q_{lime} value that equals the service life of the structure. The lime amount q_{lime} is determinable through the given permissible leaching coefficient k_1 known from experimental result analysis [2] and equal to 0.10:

$$q_{\text{lime}} = k_l V_c V_{\text{CaO}}, \tag{3.3}$$

where V_c is the cement content in the concrete close to $0.3\ \text{g/cm}^3$ and V_{CaO} is CaO proportion in cement equal to 0.65 for the Portland cement. Equation (3.3) then yields $q_{\text{lime}} = 0.0195\ \text{g/cm}^3$. In the assumption that the lime concentration will attain saturation $C_{\text{lime}} = 1.2 \cdot 10^{-3}\ \text{g/cm}^3$ long before leaching off 10 % CaO even under slow filtration, we obtain:

$$\tau = \frac{q_{\text{lime}}}{V_{\text{vol}} \cdot C_{\text{lime}}} \tag{3.4}$$

Figure 3.12 shows a plot of the function (3.4).

Since

$$V_{\text{vol}} = K_{\text{F}} \cdot \Delta H, \tag{3.5}$$

where $\Delta H = H/L$, H is the water column height and L is the structure's wall thickness, only a filtration rate as small as $v_{\text{vol}} \leq 5.15 \cdot 10^{-7}\ \text{cm}^3/(\text{cm}^2\ \text{s})$ can ensure a service life of 100 years or more at $H = 10\,\text{m}$ and wall thickness $L = 0.5\,\text{m}$.

Such estimations, although approximate, give an opportunity for understanding the requirements of a concrete's water permeability for typical structures.

In order to control concrete softening of the first kind, manufacturers introduce hydraulic additives into cements. These additives bind $Ca(OH)_2$ into less soluble calcium hydrosilicate, which dramatically reduces free lime content and suppresses its leaching rate, other factors being equal.

Corrosive Injury of the Second Kind

The concrete softening mechanism of the second kind concerns exchange reactions between environmentally occurring acids or salts and the constituents of the cement stone. The distinctive feature of concrete injury under this kind of corrosion is as follows. The concrete suffers injury only on those surface layers in contact with the corrosive media. The corrosive attack causes formation of blisters and cracks in these layers only, with all elements of the cement stone in bulk being completely preserved.

In absence of some kind of 'healing', the surface layers crumble, and deeper concrete layers suffer injury until the serviceability of the whole structure is lost.

Carbon Dioxide Injury of Concretes

Carbon dioxide (CO_2) often presents in natural water [17] and abounds in air near centers of human activity. The sources of CO_2 production in natural water are biological and microbiological processes running in water as well as in humid soils at various depths. CO_2 also evolves due to interaction of the subsurface carbonate rocks with underground waters.

At the initial stage of cement stone dissolution, carbonic acid reacts with $Ca(OH)_2$ forming calcium carbonate as the product:

$$Ca(OH)_2 + CO_2 = CaCO_3 + H_2O. \tag{3.6}$$

This process at first has a positive influence on concrete because easily soluble $Ca(OH)_2$ transforms into poorly soluble $CaCO_3$, decreasing the damageability of the concrete. Namely, the water containing carbonic acid and ions H^+, HCO_3^- and CO_3^- in equilibrium will hardly disturb the carbonate constituents of concrete. However, when CO_2 content rises over the equilibrium, the water will solve the calcium carbonate with formation of bicarbonate: $CaCO_3 + CO_2 + H_2O = Ca(HCO_3)_2$ [2], [13], which is highly soluble and easily escapes the cement stone with water.

Following the failure of cement stone, particles of an aggregate emerge on a concrete's surface. If the material of these particles is resistant to aqueous carbonic acid, they hinder penetration of the corrosive solution to deeper concrete layers but erosion of the cement stone continues in interlayers between aggregate particles.

Along the direction of filtration of acid water through the concrete, the following stages of second-order corrosion in different zones are distinguishable (Fig. 3.13):

 I. Slow calcium carbonate dissolution, its transfer through water filtration direction and, as a result, destruction of the cement stone;
 II. Saturation of calcium bicarbonate solution with oncoming calcium hydrate and precipitation of calcium carbonate consolidating the concrete microstructure;
III. Cement stone dissolution with filtered water that loses acidity in reactions during stages II and I. This stage corresponds to the concrete leaching zone.

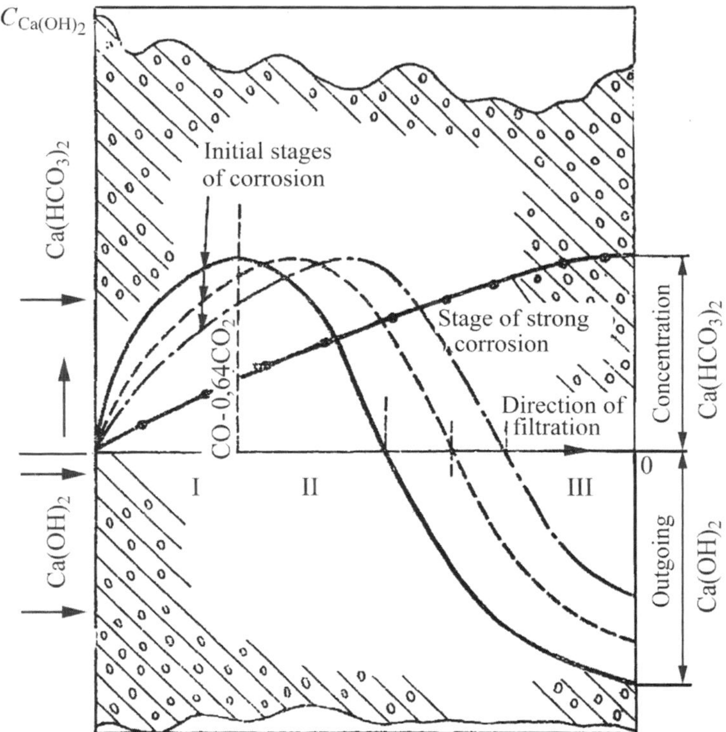

Fig. 3.13 Scheme of progressive corrosion during acid water filtration through concrete:
(I) destruction zone; (II) consolidation zone; (III) leaching zone

Fast mechanical destruction of the concrete begins when the boundary of zone (I) approaches the opposite pressure-free concrete surface, i.e., when the consolidation zone retarding the filtration disappears (is leached off).

The mechanism of concrete acid corrosion similar to the above, considered for carbonic acid as an example, is typical with minor distinctions for other acids: hydrochloric, sulfuric, oxalic, etc. Free acids rarely occur in the environment except for the above carbonic acid forming from carbon dioxide and sometimes sulfuric or sulfurous acids present in peat water [13], [14].

The second kind of concrete corrosion is most apparent in sewage constructions: sewers, collectors, waste basins, etc. Acids and bases, sulfates, nitrites, chlorides of different origin dissolved in wastewater react with cement stone as well as aggregate materials (crushed rocks) and destroy them (Figs. 3.14 and 3.15 of Appendix).

Besides, the design of sewage-related structures sometimes ignores the possible occasional dumping of emergent discharges. Concrete pipes and collectors suffer injury from such acid media and begin to pass acid or salt water into soil. Such underground water, in turn, destroys the foundations of buildings and other underground structures lying in its way, even far from the pollution source. The corrosive

Fig. 3.14 a Sewer in Zürich, Switzerland. **b** Appearance of damaged lower wall. **c, d** Disruption of concrete matrix

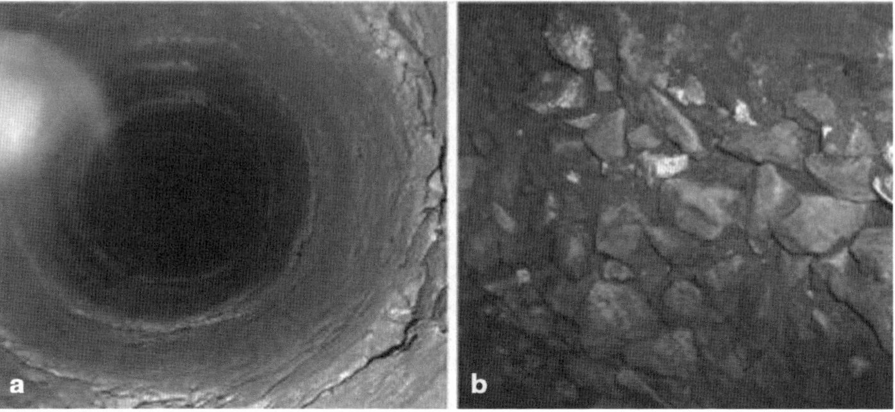

Fig. 3.15 a Sewer under Mechnikova Str. in Lviv, Ukraine, with pit and cracking damages. **b** Manifestations of concrete surface disruption due to the first kind of corrosion

injuries in concrete can be so severe that the reinforced concrete structures deform or even break down under nominal applied loads. The book [2] describes a case of the breakdown of reinforced concrete columns in the electrolysis bath shop causing ceiling collapse in an entire building. The local concrete strength degradation, or softening, in connection with the second kind of corrosion is one of main sources of crack nucleation in elements of reinforced concrete structures.

Concrete Softening of the Third Kind

The mechanism of concrete softening due to the third kind of corrosion consists of salt accumulation in pores or capillaries with subsequent crystallization resulting in an increase in the solid phase volume in relation to the solution volume. These salts are either products of chemical reactions between environments and cement stone constituents or foreign impurities carried in by mineral water. Due to a growing number and size of salt crystals in capillaries and voids, the concrete consolidates and swells. In the case of slow salt accumulation, the concrete strength can temporarily increase. Such a situation is similar to the second kind of corrosion in an aqueous solution with concentration NaOH below 10 % but differs in that the further salt content growth leads to destruction of the cement stone constituents, concrete injury, and a rapid drop in strength for the whole structure.

In particular, such transformations of cement stone can occur in water containing sulfates, most usually the dissolved calcium sulfate (gypsum). Gypsum precipitating from a solution segregates in the pores of cement stone. If the solution contains sulfates of other metals, gypsum forms in the reaction of these sulfates to $Ca(OH)_2$. Gypseous corrosion starts at concentrations of SO_4^{2-} ions higher than 1.0 g/l. The highest permissible SO_4^{2-} concentration for Portland cement is 0.25 g/l independent of the soluble sulfates composition [14].

The duration of the mechanical improvement period at the initial stage of the third kind of corrosion depends on a concrete's density and strength. For highly fluid permissible low-strength concretes, apparent destruction in the corrosive medium begins after several weeks or months.

Dense strong persistent concretes can survive several years before softening. Delayed destruction also takes place in highly porous lightweight concretes with porous aggregates because high pore volume requires accumulation of great amounts of salt, prolonging the process. In this regard, estimations of conditions, nature, and staging of destruction of a concrete by the third kind of corrosion mechanism should rely on long-term rather than short-term test results.

When inspecting concrete structures subject to renewal using injection technologies, one should remember that corrosion per se very seldom occurs. Environmental factors, as a rule, initiate local destruction and injury in concrete including contributions by all three kinds of corrosion.

The consequences of gas, physical and biological corrosion are very common and involve concrete destruction as well. Instances include breakdown of balconies (Fig. 3.16a of Appendix); column-supported sheds (Fig. 3.16b of Appendix); columns per se (Figs. 3.16c, d of Appendix), etc. Balconies and sheds break down under the influence of physical mechanisms, freezing and thawing of wet concrete, with certain contributions by gas and biological corrosion. The breakdown of vertical columns initially originates in many cases from the gas and biological corrosion of cement plasters. Further destruction accelerates due to water penetration and accumulation in pores and cracks.

Depending on the concrete softening mechanism, surfaces of injured structural elements may need specific preparation before restoration will work.

3.4 Gas Effect on Injury of Concretes

Concrete softening in gas environments, in fact, differs insignificantly from corrosions in aggressive liquids because reactions between acid gases and cement stone proceed through films of moisture where dissolved gases form the corrosive liquids.

An acid gas atmosphere is aggressive to all the minerals of cement stone. The most common of the aggressive gases is the carbon dioxide CO_2. Elevated content of CO_2 in the air can be found in industrial regions, especially in workrooms, where other acid gases, vapors, and acid aerosols can present as well.

The degree of aggressiveness in a gas medium depends on the gas's nature, its concentration, and the humidity in the air. The service life of concrete in a gas medium also depends on the integrity of the reinforcement in the structures. Therefore, predictions of durability primarily involve the corrosion resistance of protective layers, i.e., density of the cement stone.

Classification of gases according to mechanisms of aggressive action on concrete includes the following groups.

Fig. 3.16 Breakdown of balconies (**a**); column-supported sheds (**b**); columns per se (**c**, **d**) as a consequence of gas, physical and biological corrosion

The first group consists of gases that provide formation of insoluble or poorly soluble calcium salts such as CO_2, anhydrous hydrogen fluoride, silicon tetrafluoride, and fluoric anhydride. Concrete interaction with these gases results in solid phase volume increase and respective cement stone permeability decrease. Penetration of the calcium salts into bulk concrete is negligible due to small solubility. High air humidity reduces the strength of the concrete. The main cause of deterioration of reinforced concrete structures in such environments is the corrosion of reinforcement after the neutralization of protective concrete layers.

The second group comprises gases that provide formation of poorly soluble salts capturing large volumes of the crystallization water. The crystallization water significantly increases the solid phase volume and consolidates concrete as above. Such gases as sulfuric or sulfurous anhydrides and hydrogen sulfide belong to this group. For example, gypsum occupies by a volume 2.2 times larger than preceded $Ca(OH)_2$ before reaction with SO_2. Concrete density increases due to such reactions blocking deeper penetration of the gas into the concrete mass. However, the rise of high internal stresses can cause concrete delamination, in particular, peeling off of the protective layer, with subsequent reinforcement corrosion. Such phenomenon is, hence, especially dangerous.

The third group covers gases that provide formation of highly soluble hygroscopic salts absorbing water vapors and making solutions. These solutions can penetrate into the concrete by means of diffusion and/or capillary suction. They dissolve and crystallize calcium salts that, again, ensure and increase in solid phase volume and consolidation of surface layers. Nevertheless, afterwards strength decreases, and the concrete breaks down.

Gases of this group form two subgroups, according to the aggressiveness of formed salts against the reinforcement. Subgroup "a" contains gases that form salt solutions able to corrode steel and concrete even in small concentrations. These are halogen-containing gases, e.g., hydrogen chloride, chlorine. Subgroup "b" includes gases that weakly corrode steel in the basic environment of concrete, such as nitric oxides, vapors of nitric acid, etc.

Gas corrosion per se is concrete corrosion without the apparent participation of a liquid phase. In general, it is the process leading to the concrete's loss of protective function in relation to steel reinforcement. Under the combined action of acid gases belonging to different groups, carbon dioxide plays a key role in neutralization of the concrete. Other gases only insignificantly accelerate or decelerate this process. At normal air humidity and absence of condensation, the effect of these gases is negligible. However, high air humidity and, especially, moisture condensation in such cases can cause both the destruction of the protective layer and accelerated migration of good soluble calcium salts from products of cement stone decomposition into the concrete.

Therefore, in the presence of moisture condensation on the concrete surface and/or pouring the concrete structures with aggressive solutions, the structure surface requires protection through chemically resistant impermeable coatings.

3.5   Biological Factors of Concrete Injury

Biological factors of concrete injury are those destructive processes and negative changes in material or structural properties resulting from the harmful vital activity of higher or lower organisms. The effect of these factors usually depends on the products of microorganism vital activity represented by mineral and/or organic acids. Colonies of microorganisms settle on the surface of underground concrete structures. Reducing or oxidizing sulfurous compounds, depending on temperature and aeration conditions, they cause either hydrosulphuric or sulfuric acid corrosion. The same processes run on the surface of underwater concrete structures during fouling by plants and other living organisms.

Similar conditions form in industrial structures, when organic residues serving food for microorganisms accumulate on surface layers of concrete.

Since concrete is a capillary porous material, microorganisms such as bacteria, fungi, lichens, and mosses can easily settle on its surface and then penetrate into its depths, causing corrosive processes through the products of vital activity. These products include acids, sulfides, ammonia, and other agents aggressive towards concrete

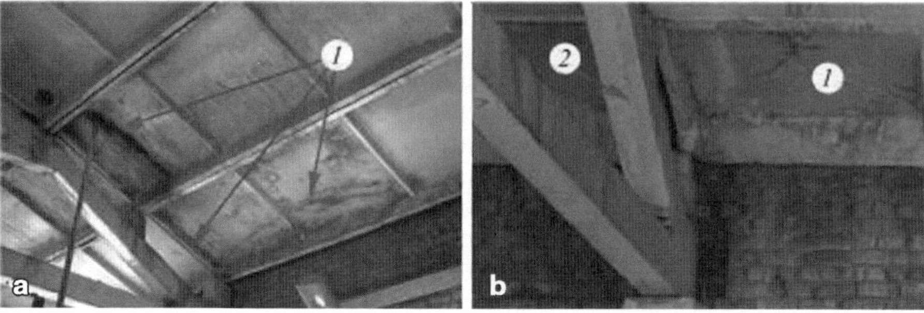

Fig. 3.17 a, b Inclined cracks *1* up to 8 m in length and 40 mm in opening in a boiling shop ceiling. **b** Reduction in ceiling panel ridge supporting depth due to concrete crumbling and column warping *2* in the frame plane

and/or reinforcement and able to destroy reinforced structures. During hardening, a thin protective film of calcium carbonate forms on a concrete's surface. Until the film is continuous, it prevents water penetration into the depth of the concrete and protects it from destruction. However, as soon as thiobacteria appear on the film surface, they begin to breed and break the film, causing continuity loss.

Biocorrosion predominantly becomes apparent as surface layer expansion or delamination occurs in concretes or, more often, moldered plasters, which gain a distinctive gamma color.

For this reason, a stereotype exists relating biological damage exclusively to the destruction of exterior concrete layers and decorative function loss. However, bacteria can penetrate deep into concrete structures. For example, improper monolithing of joints between ceiling edges and supporting columns can result in the nucleation of cracks up to 8 m in length and 40 mm in opening (Fig. 3.17 of Appendix).

The biological damaging of concretes develops most intensively under favorable conditions such as elevated air humidity and temperature. For instance, the bad condition of a boiling shop ceiling and the wetting of inner surfaces of supporting structures and wall panels can lead to crumbling of the wall joint and window frame packing, displacement of wall panels, corrosion of supporting elements, etc. (Figs. 3.18 and 3.19 of Appendix).

Removal of damaged material down to the solid surface, disinfection of the exposed surface by biocides, and recovery of the removed concrete layer in the injured places must precede repair of the reinforced structure by injection technology methods.

3.6 Damaging the Concrete Reinforcement

As commonly known [14–16], the service life of reinforced concrete structures strongly depends on the cement stone's ability to protect the steel reinforcement against various damages and injuries, especially corrosion, determined, in turn, by the presence of moisture and air in the concrete's porosity and, hence, on the steel surface.

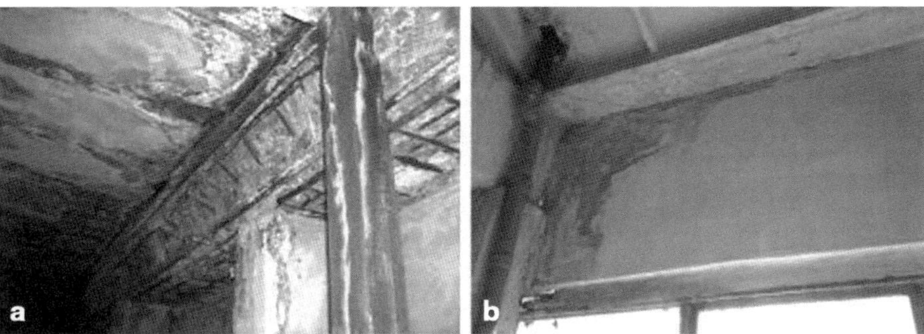

Fig. 3.18 Biological damaging of support beam, wall and ceiling panels (**a**) and wall joint (**b**) of a boiling shop accompanied by displacement of the support beam and column axes

Fig. 3.19 Vertical crack 500 mm in length and 2 mm in opening in wall supporting column of boiling shop. Bio damaging of cladding window wall panel

V.M. Moskvin [1] was the first to undertake systematic studies of reinforcement corrosion in the mid -1930s. Even the initial observation of long-term behavior of reinforced concrete structures [3] revealed that reinforcement corrosion is possible not only after destruction of the concrete's protective layer but also in the presence of such a layer, i.e., beneath it. Origin of the destruction of the concrete's protective layer may be merely mechanical, stemming from products of steel corrosion, which form with greatly relative increment of volume compared with the initial volume of steel (roughly by four times).

Reinforcement Damage by Corrosion Pits Under Static or Cyclic Loads

The static load raises both general and local rates of reinforcement corrosion, causing damage near the corrosion pit. In the presence of tensile stresses, such damages can

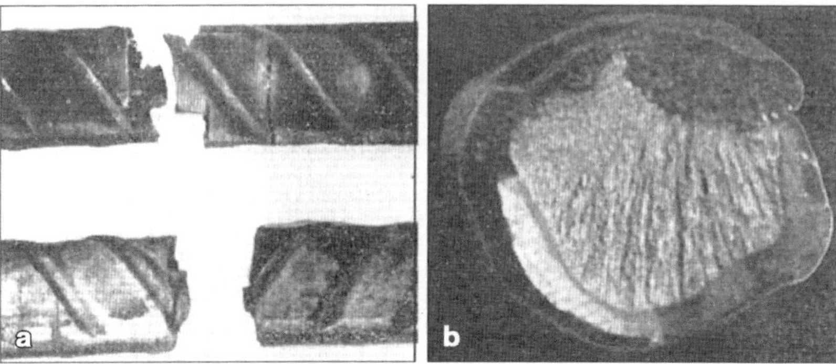

Fig. 3.20 Corrosion cracks in reinforcement (**a**) and appearance of fracture surface (**b**) at corrosion cracking [16]

cause cracks. In the presence of a respective electrochemical situation near the crack tip, such a crack can begin to grow [18].

The cyclic load is the main cause of crack nucleation in the reinforcement, even in the absence of general reinforcement corrosion [17]. A passivation film on the reinforcement surface near geometrical or technological stress concentrators breaks down under cyclic loading due to nucleation of the slip bands or surpassing the strain limit of the film. The fresh surface of metal exposed after film breakage has an anodic potential relative to the passivized surface, which assists dissolution of the metal in the slip bands and formation of new stress microconcentrators. At further cyclic loading, the probability of passivation film breakage rises, local damages grow in depth, and, finally, macroconcentrators arise with subsequent transformation into cracks.

The reinforcement fracture under static, cyclic, or mostly combined static and cyclic loads is representative (Fig. 3.20).

Dense intact concrete prepared using straight cement ensures intactness of the steel reinforcement for a long time in environments of any humidity. However, the concrete matrix degrades in the course of long-term service and crack-like defects nucleate in the material, becoming easy channels for access of environmental conditions corrosive to the steel reinforcement surface and causing corrosive injury to the mechanical steel [15], [19].

If gases or liquids have access to the reinforcement surface through cracks, then the reinforcement corrosion injury proceeds freely and self-catalytically since the corrosive media destruct the concrete matrix and widen the cracks, thus progressively accelerating the process.

Experimental studies show [15] that the crack opening exerts an essential effect on the reinforcement injury, all other factors being equal (Fig. 3.21).

The above-presented data (Fig. 3.21) were obtained by wetting specimens with water over a period of 8 h and drying them over a period of 16 h, with 150 cycles of each. The authors [13] explain that lower reinforcement injury in covered specimens occurs through the impediment of oxygen diffusion to the steel surface.

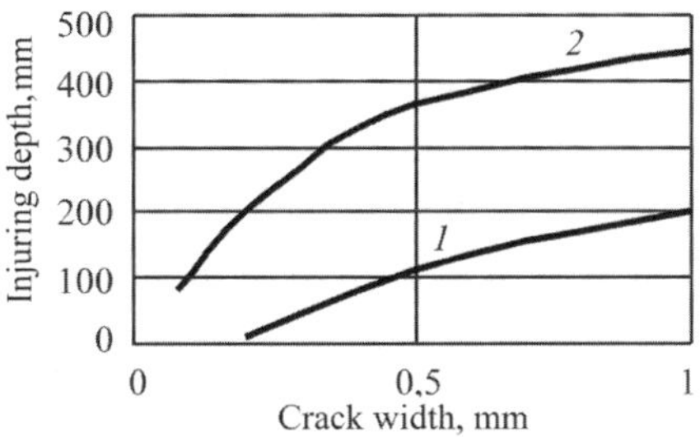

Fig. 3.21 The dependence of reinforce-ment injury depth on crack opening in concrete: (*1*) specimens covered by epoxy resin ED-5; (*2*) uncovered specimens

We can note here that the steel-reinforcement injury rate reduces over the time of corrosive environment action at all values of the crack opening in the concrete. We can explain this by filling open cracks showing products of corrosion, preventing further access of corrosive agents to the reinforcement surface.

A majority of structure failures are the consequence of steel reinforcement corrosion. Steel corrosion products formed on the reinforcement surface with a great increment of volume induce spalling of the concrete layer from the reinforcing grate with the resulting acceleration of corrosion and causing failure of the self-catalytic reinforced concrete structure.

Reinforcement corrosion injures ceiling panels in shops with aggressive gases and/or vapors (Fig. 3.22a in Appendix); support beams (Figs. 3.22b, d in Appendix), columns (Fig. 3.22e, in Appendix), etc.

Reinforcement corrosion beneath a concrete protective layer was the cause of the exterior surface injury to the cooling tower during operation (Fig. 3.23 in Appendix).

Figure 3.24 of Appendix illustrates the results of the combined effects of reinforcement and concrete corrosion under systematic loading, using the supporting framework of the seating in the Olympic stadium in Berlin as an example.

3.7 Combined Effect of Mechanical Loading and Corrosive Environments on Injury and Fracture of Concrete and Reinforced Concrete Structures

As proved experimentally [2], [7], [8], [10], [13], [20]–[23], the compressive and/or tensile stresses are the main factor determining the injury of reinforced concrete structures. In particular, trials carried out at the R&D Institute for Reinforced Concrete (Ukraine) established that the effective diffusion constant for carbon dioxide in non-reinforced concrete specimens with w/c ratio 0.47 subjected to compressive stress $\sigma \approx 0.7R_b$ decreases by 10 times, while in the same specimens subjected to tensile stress $\sigma \approx 0.7R_{bt}$ it increases by 10–100 times.

Unstressed concrete specimens with w/c ratios of 0.4 or 0.6 saturated with sodium chloride or sodium sulfide aqueous solutions (concentrations 4.7 wt. % or 5.0 wt. %, respectively) had revealed through porosity, respectively, 0.4 % or 0.56 %. Under compression up to $\sigma = 0.6R_b$ the through porosity had diminished down to 0.34 %

Fig. 3.22 Typical examples of concrete structures failure due to reinforcement corrosion injuring

Fig. 3.23 a General appearance of cooling tower with corroded reinforcement. **b** Horizontal reinforcement volutes caused by chipping of concrete protective layer (*arrows*)

Fig. 3.24 Appearance of broken sections of supporting framework (**a**) and supporting wall (**b**) in the Olympic stadium in Berlin, Germany

or 0.50 %, respectively. Under further compression up to $\sigma = 0.8R_b$ the porosity (including microcracking) had increased up to, respectively, 0.35 % or 0.56 %. The effective diffusion constants for sodium chloride or sodium sulfide solutions exhibited the same dependences on the compressive stress level. Namely, these constants decreased under compression up to $\sigma_c = 0.6R_b$ and increased at higher compressive stresses.

Under tension up to $\sigma = 0.6R_{bt}$ prism specimens of dense concrete with w/c ratio of $0.4\ldots0.6$ absorb sulfates by a rate two times more than the same specimens under compression, whereas specimens of less dense concrete with w/c ratio of $0.6\ldots0.7$ absorb sulfates by a rate three to four times more. Accordingly, sulfates penetrate deeper into the tensed concrete. Sulfate content at a depth of 2 mm was 3 times higher and at a depth of 1 mm 2 times higher in tensed concrete as compared with unstressed material. The service life of concretes in sulfate solutions was significantly longer under compression up to $0.4\ldots0.6R_b$ than the same without loading, but under compression up to $\sigma = 0.8R_b$ became less than at $\sigma = (0.4\ldots0.6)R_b$. The service life of tensed specimens at all stress levels was shorter, independent of the mineral composition of the cement.

Let us estimate the stress state effect on the concrete corrosion rate starting from the assumption that the corrosion rate is, in first approximation, governed by variations in the filtration coefficient determined by an aggressive medium penetration rate into the corrosion zone through microcracks and pores widened under tensile stresses.

In assuming that the concrete volume under tension increases only due to widening of the microcracks, the authors [2] derived an expression, which associates the mean water flow rate V_c in cracks with the concrete strain ε_t:

$$V_c = \frac{\delta^2}{3\eta} \cdot \frac{dP}{dx}\left[1 + \varepsilon_t \frac{(1 - \nu_0)}{\beta}\right]^3, \qquad (3.7)$$

where δ is crack opening (width); η is liquid viscosity; $\beta = 2\delta bn/a^2$ is the area of n capillaries with mean section $2\delta b$ each rated to total cross-section area a^2; ν_0 is the initial Poisson ratio of the concrete.

Using the above expression, the authors [2] showed that the concrete corrosion rate under tension is as follows:

$$\zeta = \zeta_0\left[1 + \frac{\sigma(1 - \nu_0)}{E\beta}\right]^3, \qquad (3.8)$$

where ξ_0 is the corrosion rate in unloaded concrete; σ is stress, and E is the Young modulus of the concrete matrix.

Under compression, dependence of the corrosion rate on stress is more complicated. In the initial period of loading from zero to crack formation stress R_T^0, when the Poisson ration of the concrete is generally constant ($\nu = 0.16\ldots0.2$), filtration and the corresponding corrosion rate gradually decrease with compressive stress growth due to tightening of capillaries, pores, and other microstructure defects. For this period, estimation of the corrosion rate is similar to above [2]:

$$\zeta = \zeta_0\left[1 + k_2\frac{\sigma(1 - \nu_0)}{E\beta}\right]^3, \qquad (3.9)$$

where $k_2 < 1$ is a coefficient accounting for closure of filtration channels and pores.

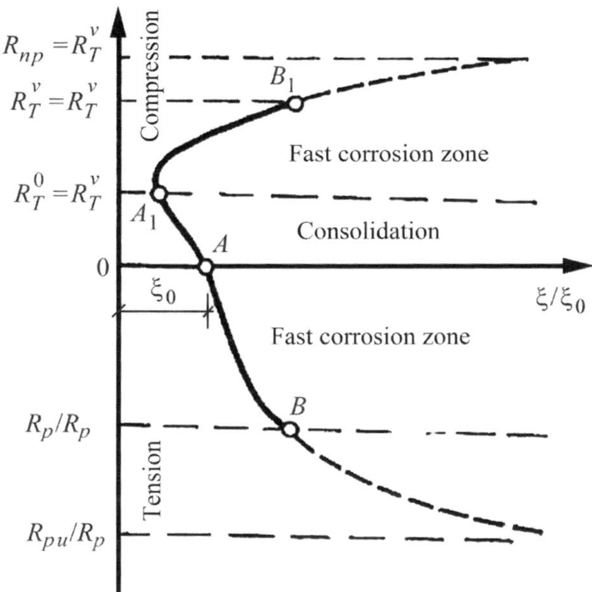

Fig. 3.25 Stress state effect on concrete corrosion rate

At higher compressive stresses $\sigma > R_T^0$ the Poisson ration of the concrete increases in accordance with the following approximate formula [2]:

$$\nu(\sigma) = 0.5 + 0.6\left(\frac{\sigma}{R_T^0} - 1\right). \tag{3.10}$$

Now, combining Eqs. (3.8)–(3.10), we can derive the following approximate expression for the corrosion rate:

$$\zeta = \zeta_0\left\{1 - k_2\frac{\sigma}{E\beta}[1 - 2\nu + \nu^2]\right\}^3. \tag{3.11}$$

The stress level effect on the corrosion rate of the reinforcement is qualitatively obvious for engineering practice from the following plot [2] (Fig. 3.25).

In section *AB,* the corrosion rate increases with tensile stress magnitude in accordance with Eq. (3.8). Conversely, under compression in section AA_1, the corrosion decelerates with applied stress, as described by Eq. (3.9). Subsequent section (A_1B_1) again corresponds to corrosion acceleration, but now under compression, as predicted by Eq. (3.11).

In conclusion, the complex influence of the environment includes such factors as chemical composition; flow rate; chemical composition variation rate; temperature regimes; pre-loading concrete saturation with solutions; saturation or periodic wetting of loaded or pre-stressed reinforced concrete; environmental impact under cyclic loading, etc. The above list reflects only the most common instances of a combined effect of environmental and internal stresses on damaging and fracture of concrete structures. Therefore, prediction of environmental effects on the concrete corrosion rate under different stress states is a complicated scientific problem, and definitively effective analytical solutions have been absent in the literature up to now.

References

1. Moskvin VM (1952) Korroziya betona (The concrete corrosion). Goslitizdat, Moscow
2. Moskvin VM, Ivanov FM, Alexeev SN, Guzeyev EA (1980) Korroziya betona i zhelezobetona, metody ih zashchity (Concrete and reinforced concrete corrosion and protective methods). Stroyizdat, Moscow
3. Collins JA (1981) Failure of materials in mechanical design: analysis, prediction, prevention. Wiley, New York
4. Popov TN (1986) Zhelezobetonnyye konstruktsii, podverzhennyye deystviyu impulsnykh nagruzok (Reinforced concrete structures subjected to pulse loads). Stroyizdat, Moscow
5. Biokorroziya, biopovrezhdeniya, obrastaniya (1976) (Biocorrosion, biodamages, fouling). CNIISK, Moscow
6. Ivanov FM, Gorshin SN, Waite J et al (1984) Biopovrezhdeniya v stroitelstve (Biodamages in construction). Stroyizdat: Moscow
7. Ahverdov IN (1981) Osnovy tehnologii betona (Basics of concrete technology). Stroyizdat, Moscow
8. Bazhenov BA (1979) Ispytaniya materialov, izdeliy i konstruktsiy (Testing of materials, elements, and structures). GGU, Gorky
9. Alexandrovsky SV, Bagriy VY (1970) Polzuchest' betona pri periodicheskikh vozdeystviyakh (Concrete creep under periodic loads). Stroyizdat, Moscow
10. Berg OY (1961) Fizicheskiye osnovy prochnosti betona i zhelezobetona (Physical fundamentals of concrete and reinforced concrete strength). GSI, Moscow
11. Arutunyan NH (1952) Nekotoryye voprosy teorii polzuchesti betona (Some problems of concrete creep theory). Stroyizdat, Moscow
12. Bazhenov YM (1970) Beton pri dinamicheskom nagruzhenii (Concrete under dynamic loading). Stroyizdat, Moscow
13. Kolokol'chikova EN (1975) Dolgovechnost' stroitelnykh materialov: Beton i zhelezobeton (Durability of building materials: Concrete and reinforced concrete). Vyssh.shkola, Moscow
14. Alexeev SN, Ivanov FM, Mordy S, Shisil P (1990) Dolgovechnost' zhelezobetona v agressivnykh sredakh (Durability of reinforced concrete in aggressive environments). Stroyizdat, Moscow
15. Moskvin VM, Alexeev SA, Verbitskiy GP, Novgorodskiy VI (1971) Treshchiny v betone i korroziya armatury (Cracks in concrete and corrosion of reinforcement). Stroyizdat, Moscow
16. Baikov VN, Sigalov EE (1984) Zhelezobetonnyye konstruktzii (Reinforced concrete structures). Stroyizdat, Moscow
17. Grunau EB (1973) Verhinderung von Bauschaden (Prevention of defects in building structures). Muller, Koln-Braunsfeld
18. Panasyuk VV, Ratich LV, Dmitrakh IM (1983) Dependence of fatigue crack growth rate in aqueous corrosive environment on electrochemical conditions in the crack tip. Phys Chem Mech Mater 4:33–37
19. Alexeev SK (1973) Korroziya i zashchita armatury v betone (Corrosion and reinforcement protection in concrete). Stroyizdat, Moscow
20. Trapeznikov LP (1986) Temperaturnaya treshchinostoikost' massivnykh betonnykh sooruzheniy (Temperature fracture toughness of bulk concrete structures). Energoatomizdat, Moscow
21. Gvozdev AA (ed) (1978) Prochnost', strukturnyye izmeneniya i deformatsii betona (Strength, microstructure transformations, and deformations of concrete). Stroyizdat, Moscow

22. Grushko IM, Illyin AG, Chikhladze ED (1986) Povysheniye prochnosti i vynoslivosti betona (Advance in concrete strength and durability). Vyssh.shkola, Moscow
23. Zaitsev YV (1982) Modelirovaniye deformatsiy i prochnosti betona metodami mekhaniki razrusheniya (The concrete deformations and strength simulation using fracture mechanics methods). Stroyizdat, Moscow

Chapter 4
Implementation of Injection Technologies for the Renewal and Restoration of Serviceability of Concrete or Reinforced Concrete Structures

Abstract General description of injection technologies suitable for practical application in strength and serviceability restoration of concrete and reinforced concrete structures of long-term operation in the construction practice and municipal services is given. Technical requirements for injection technologies are determined. Key technical aspects of preparation of concrete and reinforced concrete structures damaged by cracks or other dangerous defects for implementation of injection technologies are considered. The chapter contains a description of disadvantages of feeding aqueous cement or cement/polymer suspensions into cracks or damaged zones of concrete structures and buildings. The authors substantiate a technical feasibility of implementation of technological processes for repair and restoration of degraded building structures using fluent dual-component injection polymer materials.

General principles of implementation of technological processes based on reactive polymer composition injection into cracks or defects in concrete and reinforced concrete take place in the chapter. The technologies based on the most effective polyurethane injection material application get detailed description here. The list and specifications of injection technologies suitable for practical application in strength and serviceability restoration of concrete and reinforced concrete structures are given. Processes of injection and/or waterproofing repair of inner surfaces in sewage collectors or pipelines get detailed consideration in this chapter. Technical parameters and operational principle of the mobile diagnostic-restoration complex mounted in the truck van are characterized as well as constituting particular diagnostic or processing devices, appliances and instruments.

4.1 Characterization of Injection Technologies

In accordance with domestic [1]–[3] and foreign [4]–[6] practice, the restoration of serviceability of impaired concrete and reinforced concrete structures under long-term operation and confinement of the growth of critical defects by means of injecting a new material into the damaged places is one of the most promising approaches to the prolongation of a long-term structure's service life. Injection technologies are based on the specific materials that are suitable for processing on construction sites and provide the appropriate adhesion to concrete in cracks or damages. Such repair or restoration materials include:

V. V. Panasyuk et al., *Injection Technologies for the Repair of Damaged Concrete Structures*, DOI 10.1007/978-94-007-7908-2_4,
© Springer Science+Business Media Dordrecht 2014

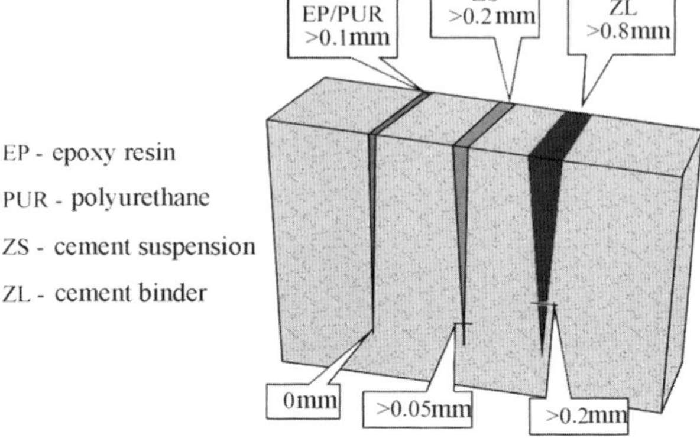

Fig. 4.1 Minimal crack openings and void portions lengths corresponding to different injection repair and restoration materials

- Aqueous paste-like suspensions of cement and mineral components;
- Aqueous paste-like suspensions of cement and polymer components;
- Fluent polymer compositions[1] of polyurethane, polyepoxy, polyacryl or silicon-organic base.

Figure 4.1 illustrates options for implementation of the most responsible injection procedure, filling cracks or defects in concrete matrices by different injection materials as proposed by the company MC-Bauchemie Müller GmbH & Co. KG (Germany). The data in Fig. 4.1 involve the following conclusions about crack opening limits and void crack portion lengths at usage of different injection materials:

- Cement binders (ZS) are applicable at crack openings over 0.8 mm and provide crack void portion length over 0.2 mm.
- Cement suspensions (ZL) are applicable at crack openings over 0.2 mm and provide crack void portion length over 0.05 mm.
- Polyurethane or polyepoxy resins (PU or EP) are applicable at crack openings over 0.1 mm and provide zero length of crack void portion.

Injection technologies for strengthening and restoring the serviceability of concrete and reinforced concrete structures must ensure accomplishment of the following technical results [7]–[12]:

- Opportunity to implement preparation of any injection compositions immediately at a construction site;
- Virtually complete filling of stress corrosion cracks down to the crack tips as well as all other defect types;
- Displacement of seepage or flowing water in cracks or defects by pressurized injection composition;
- Prevention of water with dissolved corrosive agents and microorganisms from accessing crack or defect surfaces through solidified or cured infection material;

---

[1] Fluent polymer compositions are mixtures of initial monomers, low-molecular oligomers, and high-molecular polymers.

- Interaction with concrete matrix materials involving partial (up to $1.0 \ldots 2.0$ mm) impregnation of the concrete surface with a component or injection composition itself;
- Strong adhesion bonding with concrete matrix in crack tip and edges;
- Formation of composition systems with concrete materials under repair such that they are able to withstand action of corrosive environments, mechanical loads, and/or temperature variations;
- Provision of further use of the concrete structures.

The processing of injection materials includes direct preparation at the working site at ambient temperatures. In particular, aqueous suspensions of cement or cement/polymer compositions can be prepared by mixing cement and mineral components with water in concrete mixers similar to ordinary concrete mortars. Polymer compositions are prepared by mixing two reagents: 'base' and 'hardener', or unsaturated monomer (oligomer) and initiator, which cure according to polycondensation or free radical polymerization mechanisms, respectively. To perform such mixing, we have injection pressure feeders with special reservoirs.

Injection processes are essentially distinctive for different types of repair and restorative materials. The pressurization of paste-like suspended cement or cement/polymer compositions in feeders, as opposed to polymer compositions, is impossible due to heterogeneity and the presence of mineral abrasive particles. Therefore, the common methods for concrete mortars, such as screw, membrane, or piston pumps, are applicable for feeding pastes into cracks or damages. Conversely, high-pressure feeders up to $10 \div 150$ atm are common for feeding the prepared polymer compositions into cracks or defects in concrete structures [2], [3], [7], [8].

Polymer injection materials fed into cracks and defects in the matrix of a concrete's structure cure immediately at temperatures near $20\,°C$. Aqueous cement or cement/polymer suspensions similarly solidify to common concrete mortars. Dual-component polymer systems 'base-hardeners' cure according to polycondensation or free radical polymerization mechanisms, respectively [13], [14].

4.2   Preparation of Structure for Injection

Preliminary preparation of free accessible concrete structures for injection repair or restoration work begins with as complete a removal of surface and interior layers of "weak" or loosened concrete from areas of cracking, delamination, or other defects as possible using mechanical facilities. This processing step includes the following stages and facilities:

- Removing local layers of loosened, delaminated, or cracked concrete using portable rock drills or gasoline-powered concrete saws equipped by jackbits and/or diamond disks, e.g., portable tools of the brand "Hilti" (Liechtenstein);
- Removing small portions of loosened concrete using hand tools such as hammers, chisels, boasters, and so forth.

Fig. 4.2 The mobile cleaning installation proposed by "Bodenbenden GmbH" (Germany)

Preparative works in closed sections of concrete sewage collectors or pipelines require preliminary cleaning of inner surfaces of corrosion products and dirt. Jets of dry air compressed up to 200 atm are applicable for this purpose. In addition to cleaning, such treatment provides removal of waste and water from the inner walls of the structure.

Figure 4.2 demonstrates the mobile installation proposed by "Bodenbenden GmbH" (Germany) and intended for cleaning inside surfaces of underground concrete collectors or steel pipelines with inner diameter of 250…1000 mm. In addition to the cleaning installation, the van is equipped with electric cable spool, diagnostic instruments and tools. Figure 4.3 shows the cleaning cutter in operating position.

Drilling bores for polymer composition injection Boring or drilling of operational holes $5 \div 37$ mm in diameter and $50 \div 1160$ mm in depth for feeding polymer (polyurethane) compositions is performed at pre-calculated points marked on the concrete surface. For such drilling, special boring tools designed for concrete or reinforced concrete, e.g., the tools produced by the Hilti Corp. (Liechtenstein), are used. The boring tools must be equipped with wear-resistant steel/hard alloy bits $5 \div 37$ mm in diameter and $50 \div 1160$ mm in length (see Table 4.1).

Perforator TE-15 is best suited for the following works [1], [2]:

- Drilling bores 4–25 mm in diameter and up to 550 mm in depth in concretes;
- Drilling holes in wood or metals up to 13 mm in diameter

Fig. 4.3 Mobile cleaning
cutter proposed by the
company "Schwalm
Kanalsanierung" (Germany)

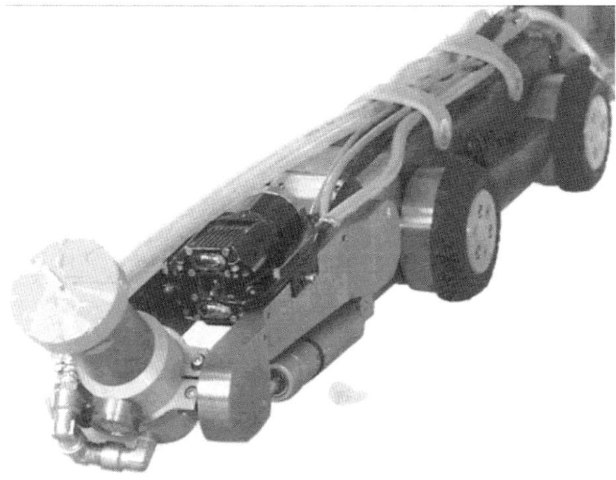

Table 4.1 General specification of boring tools produced by the Hilti Corp. (Liechtenstein)

Characteristics name	TE-15-C	TE-55
Input power, W	710	830
Impact energy, J	2.2	6.0
Rotating velocity under loading, rpm	0. . . 730	1st speed: 0. . . 230
		2nd speed: 0. . . 480
Impact frequency under loading, impacts/min	0. . . 3,850	0. . . 2,630
Recommended bore diameter range (diameter, mm)	5. . . 16	12. . . 55
Concrete boring efficiency, mm/min	370 mm/min with the bit 12 mm in diameter	220 mm/min with the bit 20 mm in diameter
Impact efficiency, mm^3/min	180,000	360,000

*The data presented for concretes of medium density

Perforator TE-55 is best suited for the following:

- Drilling bores 5. . . 37 mm in diameter in concrete or brick walls;
- Drilling bores 40. . . 55 mm in diameter in concrete or brick walls using screw augers;
- Percussion drilling bores 43.5. . . 90 mm in diameter in concrete or brick walls using impact bits;
- Chiseling bores using thin channel bits (Fig. 4.4).

Fig. 4.4 Drilling bores for feeding the pressurized injection polyurethane compositions in Lviv's concrete sewage collector at Svobody Avenue

Drilled bores for feeding the polymer (polyurethane) compositions require de-dusting by compressed air blown at pressures of $0.5 \div 2.0$ atm. After drilling, joints and cracks in the concrete require striking at depths of $100 \ldots 250$ mm.

4.3 Cement or Cement/Polymer Suspensions Feeding into Cracks and Damages

As opposed to polymer compositions, aqueous cement or cement/polymer suspensions have technological parameters unsuitable for injection into cracks or damages in concrete under pressure of the order of $10 \div 150$ atm. For this reason, such suspensions are subject to feeding into the above defects under low (below one atm) pressure using ordinary equipment for concrete mortars (see Fig. 4.5). The term 'injection' is hardly applicable to this process; it more accurately corresponds to the feeding of aqueous cement-based suspensions [14]–[16].

The process for the introduction of such materials is based on the special equipment able to carry and force thick paste-like compositions into cracks and damages.

Such machines as screws, membranes, or piston pumps are suitable for implementing the above procedures at low (below one atm) pressures (Fig. 4.5). The process rate depends on the equipment specifications presented in Table 4.2.

Unfortunately, aqueous cement or cement/polymer suspensions (concrete mortars) only partially meet the requirements for concrete injection materials, as distinct from polymer compositions. Concrete mortars suffer from serious disadvantages as materials for restoration of concrete structures since they are composed of both dissolved components and insoluble mineral particles over $100 \, \mu$m in size. The mortars have a thick paste-like consistency and can hardly fill the cracks, laminations, or other defects in concrete matrices with a width over 0.2 mm.

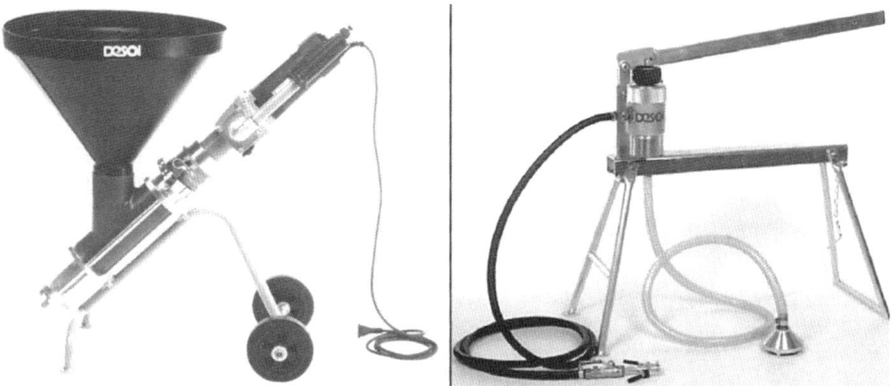

Fig. 4.5 Screw pump SP-Y (*left*) and piston pump HP-60ZD (*right*) produced by DESOI GmbH (Germany) and designed for injection materials on the cement or cement/polymer base

Table 4.2 Specification of screw pump SP-Y (*left*) and piston pump HP-60ZD (*right*) designed for injection materials on the cement base

Parameter name	SP-Y	HP-60ZD
Power, W	1800	–
Maximum working pressure, atm	15	20
Productivity	1.5... 13.5 l/min	150 ml per stroke
Highest permissible grain size, mm	3	0.3
Filling height, mm	850	–

Moreover, the parameters of the materials and features of technological processes are unable to ensure the directional delivery into the depth of the defects while the equipment is unable to raise the pressure of concrete composition over one atmosphere. Besides, the triple junction of concrete/concrete repairing material/concrete remains the place most vulnerable to long-term superimposed effects of environmental factors such as [17]–[22] in future use of the restored structures (Fig. 4.6):

- Mechanical loading (static and/or cyclic);
- Chemical and/or microbiological attacks (aqueous solutions, aggressive agents, mineral salts, and microorganisms);
- Physical influences (temperature variations, wetting/drying cycles, etc.).

Defects of the concrete structures restored using the cement-based aqueous suspensions often repeatedly crack during further use. Injection fillers introduced into the mass of the main structural material fail first (Fig. 4.7). An explanation can be found in the following:

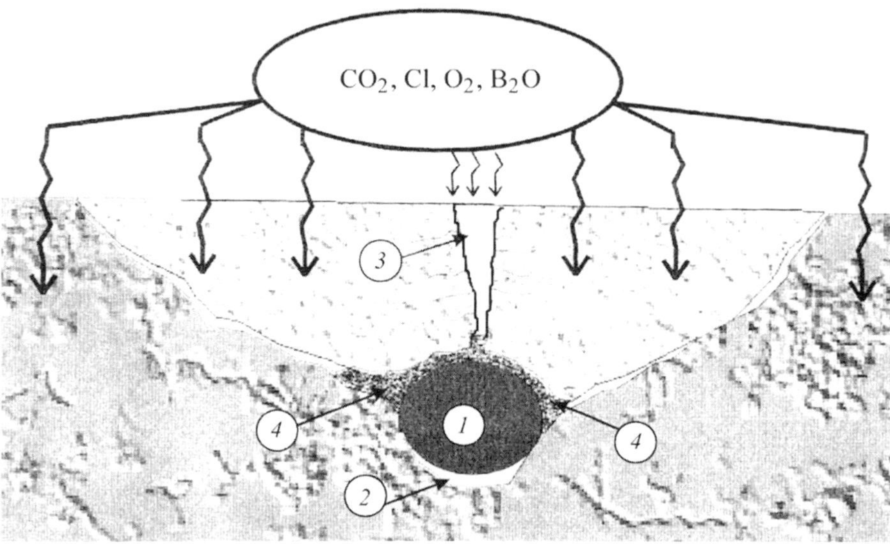

Fig. 4.6 Penetration of aqueous environments into reinforced concrete structure repaired using the concrete mortar [4]: *1* is steel reinforcement; *2* is cavity caused by concrete separation; *3* is crack; *4* are products of reinforcement corrosion

Fig. 4.7 Repeated cracking of concrete matrix defects after hardening the cement-based repairing suspension

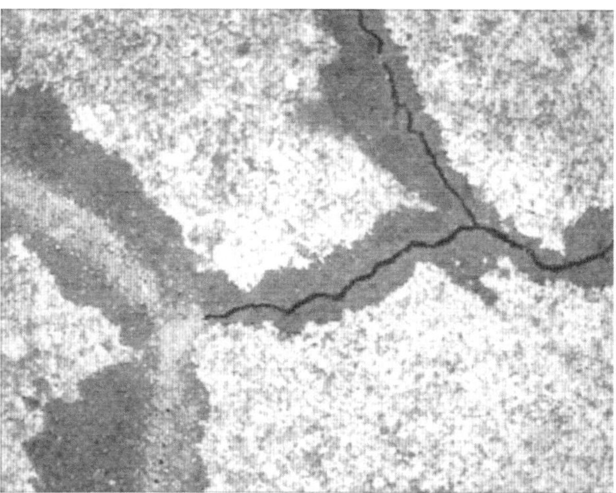

- Low adhesion of solidified repairing material to the main structural material (concrete);
- Low resistance of solidified repairing material and its adhesion junctions with the concrete matrix against mechanical (static and/or cyclic) loads;
- Low stability of the layered system in corrosive environments and temperature variations

In practice, the primary disadvantage of the aqueous cement or cement/polymer suspensions consists of incomplete filling of cracks or damages in concrete structures. The secondary disadvantage is the difficulty of handling paste-like heterogeneous repairing compositions. Besides, high content of abrasive mineral particles in the paste-like concrete compositions causes fast wear and failures of feeders and auxiliary equipment.

Application capabilities of these suspensions are limited by cracks and damages with a width over 0.2 mm. In other words, they fail to penetrate (see Fig. 4.1) into capillaries, microcracks, and microdefects typical for concrete microstructure.

Finally, service parameters of layered systems of concrete/repairing concrete/ concrete formed during repair and restoration works in macrocracks or damages are unsatisfactory [23]–[29]. All of the above considerations lead to the inevitable conclusion that the polymer injection materials are more effective and find wider implementation in many cases of effective strengthening and renewal of functional characteristics of damaged concrete structures.

4.4   Polymer Composition Injection into Cracks and Damages

Implementation of polymer material injection technologies for strength and durability renewal in concrete structures has proved the high efficiency of these technologies [1]–[3], [7]–[10][2]. The highest positive results corresponded to the case of feeding pressurized polyurethane compositions into the depth of structural concretes damaged by deep surface or inner cracks. This conclusion complies with the practical results of foreign specialists [4]–[6], [23], [28].

Nearly perfect filling of cracks, laminations, cavities or other defects with polymer compositions was attainable in both overground and underground concrete and reinforced concrete structures all over the country. The technical requirement for this operation consists of injection of polyurethane compositions into a concrete mass under pressure of $10 \div 150$ atm. Injected compositions must be able to structurize by forming solid polymers at temperatures near 20 °C [1]–[3].

Despite the common injection principle, there exists an essential difference between approaches to filling and locking of dry or wet (water-filled) cracks during operations on the strength of the concrete matrix and restoration of service life (Figs. 4.8 and 4.9). In the presence of condensed moisture or flowing water on concrete surfaces inside cracks or other defects, the technical requirements for polymer compositions and injection implementation parameters significantly increase.

First, wetting of concrete surfaces in a stress corrosion crack tip and edges negatively effects the bonding and subsequent hardening of polymer compositions on such surfaces. Second, the moisture on the concrete surface prevents the formation of

---

[2] The State Production & Research Center "Techno-Resource" of the National Academy of Sciences of Ukraine had developed the necessary domestic polymer compositions, performed in-lab tests and field trials, and implemented the technology into practice.

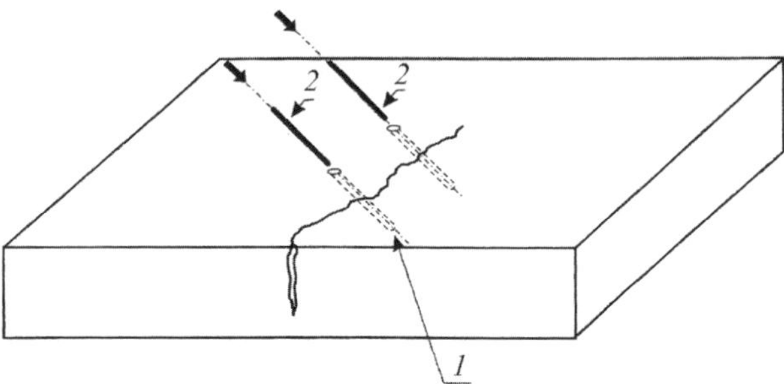

Fig. 4.8 Principal scheme of polymer composition injection into the dry crack: *1* denotes the drilled bores; *2* denotes packers for composition feeding

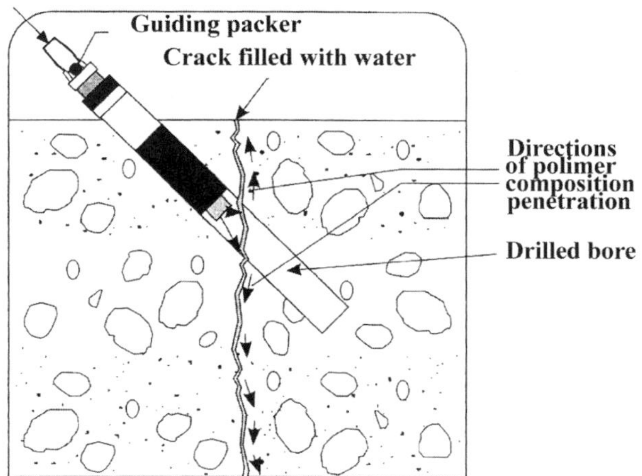

Fig. 4.9 Principal scheme of polymer composition injection into the water-filled through crack [4]

stable adhesion junctions of 'concrete/polymer/concrete'. This reduces the strength of the layered systems formed during repair and restoration work.

Theoretically speaking, the injection polymer compositions fed under pressure of $10 \div 150$ atm are able to stop water flow, displace it from stress-corrosion cracks or other defects of concrete matrices, and occupy the free space. Nevertheless, all this is insufficient for the restoration of complete concrete structure serviceability. The compositions must cure with negligible shrinkage and bond as strongly as possible to the concrete surface.

The necessary technological routes, applicable in field conditions and commercially feasible, are developed, preventing opening of the injection composition-filled crack and/or repeated water inrush, including [10], [30]–[36]:

- Fast wetting and, if possible, partial impregnation of concrete surfaces by the injection composition or its components;
- Provision of high strength and hardness of the cured solid polymer formed from the injection composition;
- Provision of strong adhesive bonds in the layered system of 'concrete/polymer/concrete'.

Other polymer compositions such as silicon-organic, polyepoxy, or polyacrylate are unable entirely to displace microdrops of moisture from crack edges or dry the wet upper surface layers of concrete [37]–[40]. Therefore, only those injection polymer compositions that are able to substitute water in concrete surfaces and/or chemically fix it with the formation of high-molecular compounds stable over long-term service are preferable in the case of wet cracks. Such materials first include two-component (polyol-polyisocyanate) polyurethane systems [30], [41]–[43].

4.5 Applications of Polyurethane Injection Materials

The polymer injection compositions that satisfy the technological requirements of fluidity and ability to fill under pressure the entire volume of inner cracks and other defects in concrete matrices (Fig. 4.1), create strong adhesive bonds, and hardening at temperatures near $+20\,^{\circ}\mathrm{C}$, comprise silicon-organic, polyepoxy, polyacrylate, polyester, and polyurethane materials [44]–[47]. The polyurethane compositions are notable among them for a combination of properties.

Both single-component and dual-component polyurethane compositions possess considerable advantages as compared with the other above-listed injection materials. Cured polyurethane elastomers have the highest stability in water environments saturated with chemical agents and microorganisms (bacteria or fungi) [45]–[48]. The stability of the elastic polyurethanes under both static and cyclic loading [49]–[55] is also of high importance. These features create the prerequisites for prolonged serviceability of the layered junctions of 'concrete/polyurethane/concrete' under stress-corrosive conditions.

According to the data of foreign specialists, polyurethane injection compositions have great technological advantages over systems based on other polymer compositions [1]–[3], [7]–[9], [41] when cured in cracks and damages of concrete structures. These advantages, in particular, reflect in the experimentally established stability of the composite systems of 'concrete/polyurethane elastomer/concrete' under high deformations of the material. For example, tensile strain of a concrete structure bound with the polyurethane binder can be as high as 5.0 %. Under the same conditions, the composite junction of 'concrete/polyepoxy/concrete' breaks at the tensile strain value of 1.0 % [4].

Authors [2]–[4] have substantiated implementation of polyurethane materials in the practice of strength and serviceability restoration of concrete structures. In comparison with silicon-organic or polyepoxy analogues, dual-component polyurethane compositions are simpler to prepare at construction sites and easier to introduce under pressure into the concrete matrices being repaired through drilled bores.

Another confirmation of the efficiency of polyurethane compositions in healing cracks and damages in concrete matrices is obvious from the fact of prolonged service life of concrete/polyurethane elastomer/concrete junctions [35]–[38]. Polish specialists have shown [4], [6], [11] that impregnation of polyurethane composition into a concrete mass under high pressures entirely restores the serviceability of the damaged structural elements [7]–[9], [51]–[55].

One more essential technological advantage of polyurethane injections vs. silicon-organic or polyepoxy injection systems is in the ability of isocyanates (second components) to cure with a great volume expansion due to capturing water molecules. Such expansion blocks the propagation of cracks or damages and stops water leakage in sewage collectors and hydraulic constructions. The studies [56], [57] have demonstrated that the ability of polyurethane compositions to react with water explains both easy feeding and formation of polyurethane foams. Note that the formation of solid polyurethane foams immediately at construction sites is much easier than foam plastics from other polymers. Besides, polyurethane foams excel in hardness, strength, and resistance against environmental impacts and aqueous aggressive media in a wider working temperature range [57], [58] when compared to other foam materials.

Polyurethane foams solidify in five minutes or less and simultaneously increase volume by $15 \div 30$ times [56]–[58]. Moreover, formed in a concrete mass, polyurethane foams, after structurizing and hardening, create strong adhesive bonds with the concrete matrix in crack tips and between opposite crack or lamination edges, thereby strengthening the damaged structure or element. Simultaneously, the solid foam with a closed cellular structure blocks ways for water penetration through cracks or damages of the concrete matrices.

The technological disadvantage of foamy polyurethane injection compositions (of $15 \div 30$-fold volume expansion) consists of much weaker adhesion to the repaired concrete matrix as compared with the compact polyurethanes. Besides, in spite of the closed cellular structure of polyurethane foams, corrosive aqueous media dangerous for concrete can penetrate them through pores over long-term use in contact with water environments [59], [60].

The main method for elimination of this disadvantage consists of additional feeding of pressurized compact polyurethane compositions unable to foam in 10...30 min after stopping water leakage or seepage through cracks or damages in the concrete matrix by injected polyurethane foams. In order to feed the compact polyurethane, the same injection feeders and guiding packers in drilled channels are used. As a result, the solid polyurethane elastomer arises inside the polyurethane foam in the system of 'concrete/polyurethane foam/concrete', raising both the strength and waterproofing parameters of the repair and restorative composite system.

The restorative process in the case of wet or water passing cracks or damages in concrete structures includes two main routes [2]:

- Introduction of the fast foam forming polyurethane material in an amount high enough for complete filling of cracks or damages and stopping of water leakage or seepage, and
- Filling voids in the polyurethane foam with compact non-foamy polyurethane under high pressure.

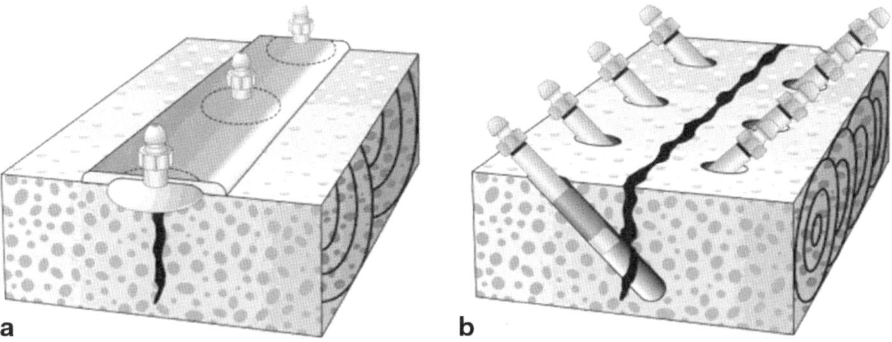

Fig. 4.10 Schematic layout of bonded (**a**) or screw-in (**b**) packers for feeding the injection polymer compositions into damaged zone

4.6 Equipment for Strengthening and Restoration of Concrete Structures Using Polyurethane Injection Materials

The process flow sheet includes the following main routes [61]–[63]:

- Visual inspection and instrumental examination of concrete damages with subsequent classification (see Chap. 6 in this book);
- Determination and marking of points for drilling bores in the damaged area for feeding injection compositions;
- Drilling bores at an angle of 45° to the structure surface with a depth of 0.5 . . . 1.0 m (or down to the middle of the concrete body), and
- Inserting packers into the drilled bores for feeding the pressurized injection compositions (Fig. 4.10).

The packers (Figs 4.11 and 4.12) not only guide the polyurethane composition in the required direction through the drilled channel, but also preserve an injection pressure created by the feeder inside the concrete matrix. This function by itself permits impregnation of the liquid polyurethane material into fine cracks, pores, and other defects.

The special technological equipment (feeders) provides injection of polymer compositions into cracks and damages in concrete structures during repair or restorative works. The plunger or membrane type feeders are most commonly used (see Tables 4.3 and 4.4). Figures 4.13 and 4.14 demonstrate certain feeders designed for feeding various strengthening polymer materials into the concrete matrices under pressure of $10 \div 150$ atm. They enable the preparation and pressurization of single-component or dual-component, high viscous, predominantly polyurethane compositions.

Pumps of various types and design deliver fluent injection compositions through high-pressure hoses to the distribution systems (manifolds) and then through packers to cracks or damages in the concrete structures. The single-component injection materials arrive under pressure directly from reservoirs connected to the pumps. The

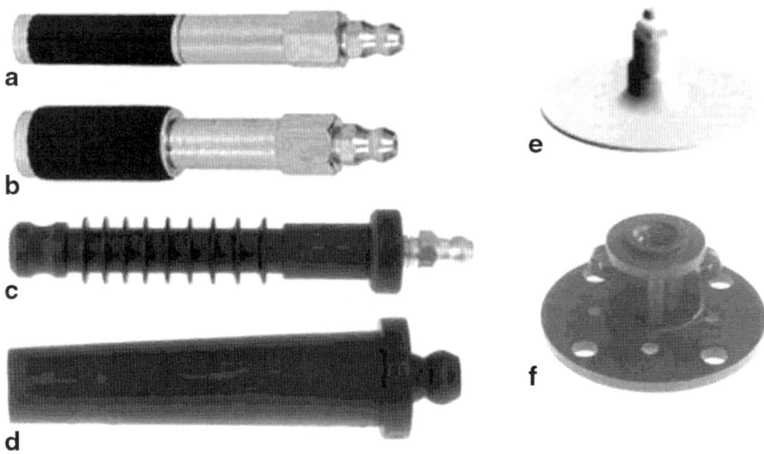

Fig. 4.11 Injection packers of different types: screw-in metal packers (*a*, *b*); hammered plastic packers (*c*, *d*); bonded plastic packers (*e*, *f*)

Fig. 4.12 Packer on concrete structure surface sealed with a polyepoxy paste [4]

two-component materials need preliminary preparation of a reactive composition using special mixing heads.

Figure 4.14 demonstrates the mixing head (gun) and its parts produced by SCHOMBURG GmbH (Germany) and designed for the preparation and delivery of injection compositions. Figure 4.15 shows the connection of another mixing head WEBAC to a packer inserted into the damaged area of a foundation.

Table 4.3 Specification of injection pumps made by WEBAC® Chemie GmbH (Germany)

Characteristics	Membrane pumps				Piston single-component pumps		
	Single-component			Dual-component	Hand HP 100	Electric drill of type HEP 1001 mounted	Hand press mounted
	IP 1	IP 2	IP 3	IP 2K -F1			
Operating pressure, atm	0…250	0…250	0…250	0…200	0…100	0…400	0…300
Productivity, l/min	1.6	2.5	5.5	14	0.07 l/cycle	0.5	1.2
Power, kW	0.55	0.75	1.8	–	–	–	–
Material delivery hose length, mm	3,000	6,000	6,000	5,000	3,000	500	–
Charging camera volume, liters	6	6	6	–	–	1	–

Table 4.4 Specification of injection feeder made by DESOI GmbH (Germany)

Pneumatic piston pump KOMPAKT PN 1412-3K

Operating pressure, atm	10...200
Productivity, l/min	8.5
Transformation ratio	1:2.5
Composition components ratio	1:1
Air pressure, atm	10
Compressor productivity, l/min	500

The restorative process of renewing damaged sections in a building (Fig. 4.15) consists of feeding the pressurized injection polyurethane (or another polymer) composition into bores drilled in the brickwork down to the damages. Components A and B (base and 'hardener') flow from the injection feeder to the mixing head through two hoses. The operator can raise the pressure of the mixed composition, if needed, by opening a reducing valve on a compressed air line. Injection of the compositions occurs through the guiding packers. Figures 4.16–4.18 illustrate an example of water leakage elimination in the reinforced concrete dam of the Tashlytska hydroelectric pumped storage power station (Pivdennoukrainsk, Mykolayiv Region, Ukraine) by the injection of fluent polyurethane compositions. The State Production & Research Center "Techno-Resource" of the National Academy of Sciences of Ukraine performed these repair and restorative works.

Figures 4.19 and 4.20 illustrate the process routes of pouring components A and B of the polyurethane composition into the feeder and injection of this composition under pressure into inner cracks, damages, or voids in the block stone basement of an old house (Lviv, Ukraine). Injection restoration not only strengthens the stone basement, but also prevents possible water penetration through cracks and damages in the cement-bonded block-stone walls (old houses in the city of Lviv are located on aqueous soil).

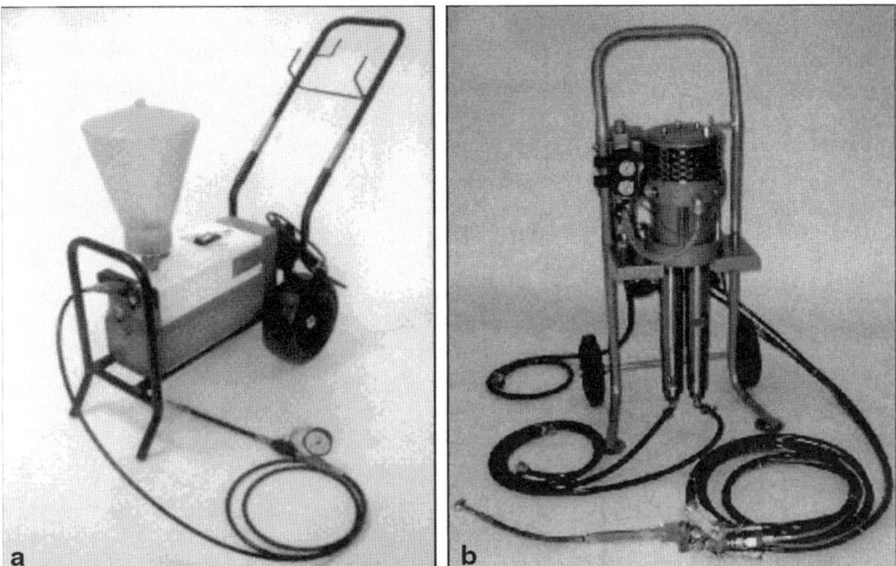

Fig. 4.13 Injection feeders made by WEBAC® Chemie GmbH (Germany) for single-component (**a**) or dual-component (**b**) polyurethane compositions

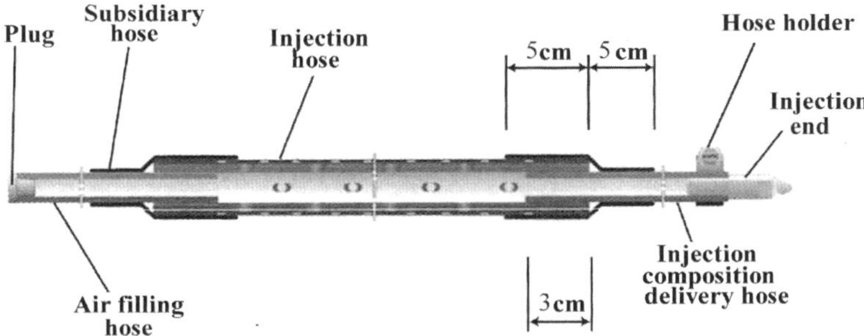

Fig. 4.14 The design of mixing head of SCHOMBURG GmbH (Germany) for cement-polymer or polymer compositions

After injection and hardening of the polyurethane compositions, the areas of concrete surface exposed through removal of the impaired concrete require finishing by sandblasting until they have reached the first surface finish class according to GOST 9.402-89 or finishing level Sa2 1/2 according to ISO 8501-1. Optionally, a steel brush treatment can be used instead of sandblasting until they have reached the second surface finish class according to GOST 9.402-89 or finishing level St2... St3 according to ISO 8501-1. The finished surfaces need protection with modern waterproofing polymer/cement compositions based on aqueous suspensions.

Fig. 4.15 Mixing and delivery head WEBAC (Germany) for two-component polymer compositions and its connection to a packer

Fig. 4.16 Surface of the reinforced concrete dam of the Tashlytska hydroelectric pumped storage power station (Pivdennoukrainsk, Ukraine) in the turbine house revealing zones of water leakage through microcracks and damages

4.7 Process and Equipment for Strengthening and Restoration of Sewage Collectors and Water Conduits Using Injection Methods

Restoration of reinforced concrete sewage collectors and water pipelines with serious damage is an urgent problem of municipal engineering [64]–[67]. Several enterprises, such as SPRC "Techno-Resource" N.A.S. Ukraine, Bodenbenden GmbH

Fig. 4.17 Employees of SPRC "Techno-Resource" N.A.S. Ukraine fill in the reservoir of the high-pressure $(50 \div 150 \text{ atm})$ injection feeder with a polyurethane composition

Fig. 4.18 Injection of pressurized single-component polyurethane composition into wet cracks and damages of the reinforced concrete dam of the Tashlytska hydroelectric pumped storage power station (Pivdennoukrainsk, Ukraine)

Fig. 4.19 Employees of SPRC "Techno-Resource" N.A.S. Ukraine perform repair and restorative works in an old house basement (Lviv, Ukraine) using polyurethane compositions and a low-pressure (below 12 atm) feeder

Fig. 4.20 Injection of a dual-component polyurethane composition through a packer into the foundation during repair and restoration works in the basement of the Hotel "Leopolis" in the city of Lviv (performed by SPRC "Techno-Resource" N.A.S. Ukraine)

(Germany), Rabmer Rohrtechnik GmbH (Austria), IBAK Helmut Hunger GmbH & Co. KG (Germany), and Tech-Kan (Poland), have achieved great success in solving this problem. The practical experience of these enterprises proves the efficiency of injection technologies in the fulfillment of real tasks using, in particular, the following processes:

- Healing cracks and/or damages using fluent polymer compositions with subsequent hardening (Fig. 4.21);
- Repairing the structures using composite systems based on glass fabric impregnated with liquid synthetic resins.

Each specialized enterprise or firm, while renewing damaged structure elements using injection technologies, often contributes its own modifications into the process implementation and develops different original equipment (Fig. 4.22). However, the mandatory process route in all such modifications related to water conduits or collectors is pre-treatment and thorough cleaning of the damaged areas. Special equipment exists for this purpose, designed to scavenge inside the pipe surface with the aid of brushes, cutters, or water jets under pressure up to 250 atm (Fig. 4.21a).

Another important common process route consists of blocking the repaired pipe section with two rubber restraining cylinders adjoined to the inner pipe surface from both sides. Then, the feeder pumps the repairing composition into the space between these cylinders through sealed inlets under pressure up to 10 atm (Fig. 4.21b). Cylinders expand in diameter and closely fit to the inner pipe surface while the polymer composition solidifies and heals the damage.

The alternative technological approach (Fig. 4.22) to restoration of serviceability of damaged collectors or water conduits depends on bringing a calibrated rubber appliance with a glass fabric reinforced patch covered by a high-viscous polymer layer to the repaired section. Then, the compressed air presses the reinforced patch to the pre-cleaned surface of concrete or steel pipe in the damaged section and the patch enters into adhesive bonding with the damaged area.

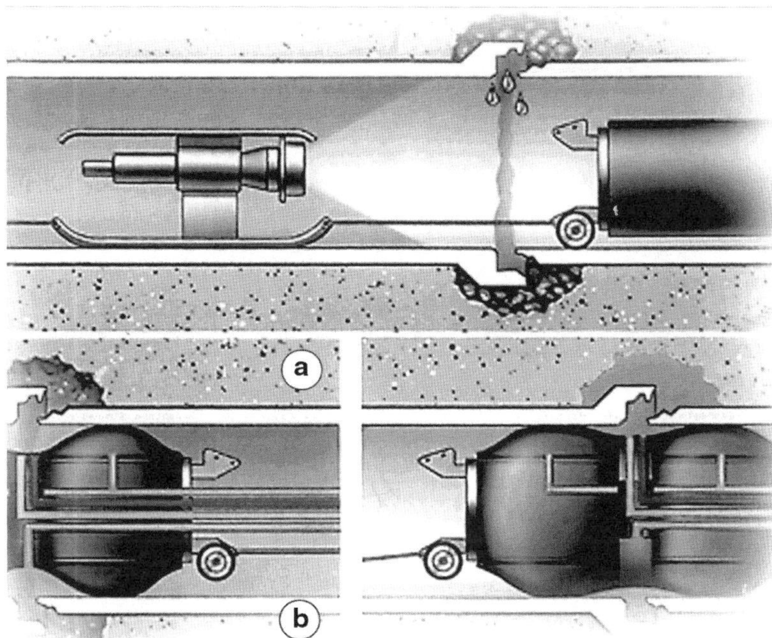

Fig. 4.21 The process flow sheet of surface cleaning and repair in damaged concrete collectors or pipelines using polymer compositions

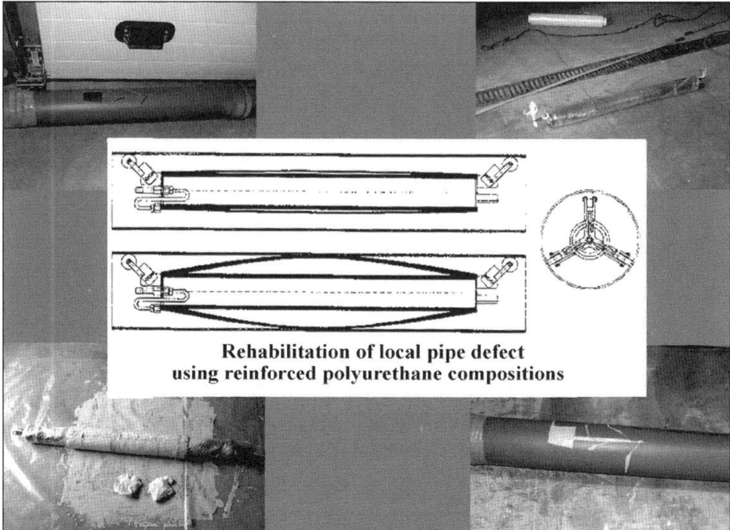

Fig. 4.22 Restoration of damaged pipeline by remote application of the nonwoven fabric impregnated with epoxy resin mixed with amine hardener

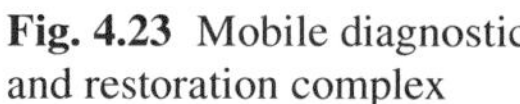

Fig. 4.23 Mobile diagnostic and restoration complex

After being pressed onto the damaged area, the patch stays under pressure during the specified time (from a few minutes to two days) to ensure complete polycondensation or polymerization of the impregnated dual-component compositions. When the polymer binder cures, the restored section of damaged concrete or steel pipeline is subject to checking of tightness, with subsequent removal of the pressing appliance if the check has yielded the positive result.

4.8   Mobile Diagnostic and Restoration Complex

The demand for efficiency and high quality in injection process routes during repair and/or restorative works in damaged building structures induces a certain necessity in the mobile diagnostic and restoration complexes. The design and manufacture of such complexes equipped with up-to-date technological facilities and measuring instruments was the aim of an innovative project carried out in 2006 by scientists of the G.V. Karpenko Physical & Mechanical Institute N.A.S. Ukraine. Employees of SPRC "Techno-Resource" N.A.S. Ukraine provided further implementation of the complexes in municipal structures of the State Committee for Construction of Ukraine, as well as the hydroelectric pumped storage power stations of the National Energy Company "Energoatom" and JSC Ukrgidroenergo. The general purpose of the complex consists of damaging diagnostics, injection strengthening, and waterproofing concrete, reinforced concrete, and steel structures comprising sewage collectors; main conduits; tunnels of underground railway; tunnels for rail and motor transport; dams and other hydraulic structures; bridges, foundations, and walls of industrial and residential buildings, etc.

In order to deliver the qualified personnel to remote building sites as well as accommodate all diagnostic instruments, technological equipment, and materials, the diagnostic and restoration complex was mounted on a five-seat motor van with a three-ton capacity (Fig. 4.23). In particular, such a van contains:

- An instrumentation system developed by SPRC "Techno-Resource" for video diagnostics SVD-1 including data transfer means and personal computer (Fig. 4.24);
- Equipment package IN-1 for preparation and injection of polyurethane compositions (Fig. 4.25);
- An equipment package REN-1 developed by SPRC "Techno-Resource" for pipeline rehabilitation (Fig. 4.26);
- Supplementary equipment including portable power plant; compressor; operator's workplace; shelving; repair toolkit; personal protection means, etc.

The installed equipment enables *in situ* diagnostics and repair of damaged structures. The purpose of the mobile system of video diagnostics SVD-1 is inspection of concrete or reinforced concrete sewage collectors and steel pipelines with a diameter under 500 mm. The system records video along 360° of the pipe's perimeter at a distance up to 50 m that allows registering size and location for all defects including cracks, cleavages, etc. Based on video monitoring and computer video records, the operator makes decisions concerning structure repair methods and estimates the quality of the finished work.

Key specifications of diagnostic and restoration equipment of the complex are as follows:

Mobile System of Video Diagnostics SVD-1

Case protection class	IP67
Overall dimensions, mm	$400 \times 115 \times 123$
Mass, kg	6
Supply voltage, V	12
Consumption current (max), mA	50/(150)
Gear ratio	188:1
Axial moment of force, mNm	140
Temperature range,°C	From -20 to $+60$
Transmission ratio	1.3:1
Wheel moment of force, mNm	182
Travelling speed, m/hr	~ 90
Sensitivity of video camera, lux	0.2
Rotation angle of video camera around horizontal axis, deg	355°
Rotation speed of video camera around horizontal axis, deg/s	30
Rotation angle of video camera around vertical axis, deg	0–90°
Rotation speed of video camera around vertical axis, deg/s	30
Image zoom	$\times 2 - \times 10$
Shutter rate (exposure), s	1/60–1/120000
Video frame resolution, dpi	>480
Signal/noise ratio, dB	>50
Video signal voltage (75 Ω), V	~ 1.0
Working temperature,°C	0…40
Video camera supply voltage, V	12
Consumption current, mA	500
Cable length, m	65
Video signal capture method	analog

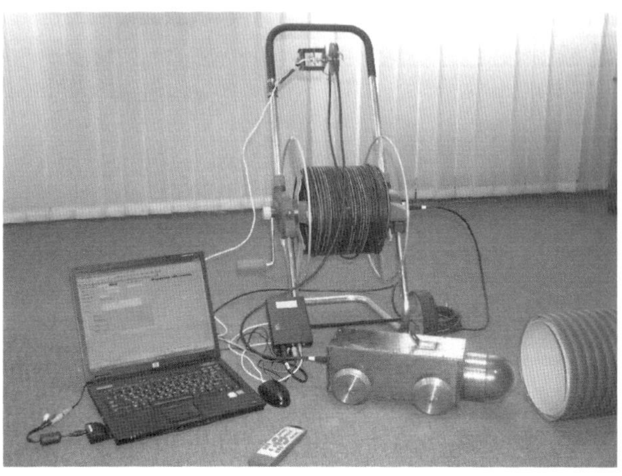

Fig. 4.24 Mobile system of video diagnostics SVD-1 with video head and computer-based system of signal capture and recording

Equipment Package IN-1 for Preparation and Injection of Polyurethane Compositions

Pressure reduction ratio	33:1
Maximum productivity at free outflow, l/min	3.0
Maximum composition feed per cycle, cm^3	14
Maximum inlet air pressure, Bar	8
Maximum operational pressure, Bar	264
Air loss (per cycle at pressure 8 Bar), liters	3.8
Maximum environmental temperature, °C	80
Overall dimensions, m	$0.14 \times 0.34 \times 0.55$
Mass, kg	9
Noise level at workplace:	
–Idle, dB	75
–loaded, dB	73

Equipment Package REN-1 for Pipeline Rehabilitation

Piston diameter, mm	100	200
Length, mm	1100	2000
Maximum pipeline diameter, mm	250	450
Maximum operational pressure, Bar	8	8
Mass, kg	2.5	7.8

The equipment package IN-1 is designed for rehabilitation of large-size building structures (dams, tunnels, bridges, underground galleries, etc.) using injection of polyurethane compositions. Before repair, the necessary engineering estimations are

Fig. 4.25 Repairing of concrete sewage collector at the Svobody Avenue, city of Lviv, Ukraine

Fig. 4.26 Equipment package REN-1 for pipeline rehabilitation

required depending on the damage level of the structures or buildings, the character and size of cracks and defects, service conditions, etc. Then, employees supply the fluent polyurethane composition under pressure of 50 ... 150 atm to the defect zones using the installation IN-1 operated by compressed air.

For small-sized building structures such as sewage collectors or pipelines with a diameter under 500 mm needed in local rehabilitation, the equipment package REN-1 is used. Its principle of operation consists of bringing a rubber piston with the synthetic fabric reinforced patch impregnated by a fluent polymer composition to the repaired section. Supplying the compressed air under pressure of up to one ... three atm inside the rubber piston causes its expansion by pressing the patch with the polymer composition to the inner surface of the pipe at the damaged place. Under such pressing, the polymer composition fills cracks and microcracks in the defect area while impregnated with the same composition patch forms reinforced covering after polymer curing. The reinforced covering strongly bonds to the inner surface of the collector or pipeline and eliminates the damage.

References

1. Panasyuk VV, Sylovanyuk VP, Marukha VI (2005) Strength of structure elements damaged by cracks and healed using injection technologies. Phys Chem Mech Mater 6:60–64
2. Marukha VI, Serednitskiy YA, Gnip I P, Sylovanyuk VP (2007) Development of injection technologies and design of mobile equipment for diagnostics and serviceability restoration of concrete or reinforced concrete structures operating under stress-corrosive conditions. Sci Innov 3:55–62
3. Sylovanyuk VP, Marukha VI, Genega BY, Ivantiskhin NA (2002) Fracture mechanics as the base of densification injection technology in rehabilitation of long-term objects. In: Mekhanika i fizika ruynuvannya budivel'nikh materialiv i konstrukciy (Mechanics and physics of fracture for building materials and structures), No. 5. Kamenyar, L'viv, p 373–382
4. Czarniecki L, Emmons PH (2002) Naprava i ochrona konstrukcji betonowych (Repair and protection of concrete structures). Polski Cement, Kraków
5. Allen RNL, Edwards SC, Shaw DN (eds) (1993) Repair of concrete structures. Blackie Academic and Profectional, Glazgow
6. Wysokowski A, Żurawska A (1998) Wprowadzenie w tematyke nowoczesnego wzmacniania mostów betonowyckh (Advances in concrete bridge strengthening). In: Nowoczesny metody wzmacniania mostów betonowych (Modern methods of concrete bridge strengthening). Instytut Badawczy Dróg i Mostów, Warsaw, p 126–129
7. Marukha A, Genega B, Serednitskiy Y, Zaplatins'kiy M (2006) Concrete structure protection against stress corrosion using polyurethane injection compositions. Phys Chem Mech Mater 5:834–840
8. Marukha VI, Genega BY (2001) Ushchil'nyuval'ni tekhnologiï— dlya zmicnennya i remontu zalizobetonnikh konstrukciy (Consolidation technologies for strengthening and repairing reinforced concrete structure). In: Diagnostika, dovgovichnist' i rekonstruktsiya mostiv i budivel'nikh konstrukciy (Diagnostics, durability, and rehabilitation of bridges and building structures). No. 4. Kamenyar, L'viv, p 158–161
9. Marukha VI, Genega BY, Serednitskiy YA (2006) Efektivnist' zastosuvannya poliure-tanovikh in' єktsiynikh materialiv u vidnovlenni pratsezdatnosti betonnikh i zalizo-betonnikh konstruktsiy i sporud z koroziyno-mekhanichnimi trishchinami (Polyurethane injection materials efficiency in serviceability renewal of concrete and reinforced concrete structures with stress-corrosion cracks). In: Diagnostika, dovgovichnist' i rekonstruktsiya mostiv i budivel'nikh konstrukciy (Diagnostics, durability, and rehabilitation of bridges and building structures). No. 8. Kamenyar, L'viv, p 84–90
10. Marukha VI, Vasilechko VO, Genega BY et al (2003) Waterproofing cover selection rules for protection of a sewage collector against very aggressive media volleys. In: Proc. int. water forum "Aqua Ukraine 2003", Ukrainian Water Association, Kyiv, Nov 4–6, 2003, p 194–195
11. Broniewski T, Ciesielski R, Fiertak M (1999) Technical condition evaluation and service life prediction for industrial reinforced concrete pipelines. Corros Prot 1:7–12
12. Velesovski RA, Kestelman VN (2004) Adhesion of Polymers. McGrawHill, Peking
13. Fakirov S (ed) (2005) Handbook of condensation thermoplastic elastomers. Wiley-VCH, Weinheim
14. Gotz VI (2003) Betony i budivel'ni rozchiny (Concretes and building mortars) KNUBA, Kyiv
15. Ivanov FM, Yakub TY, Chayka NA (1972) Vliyaniye struktury betona na yego korrozionnuyu stoykost' (Concrete microstructure effect on its corrosion resistance). In: Korroziya i zashchita stroitel'nykh konstruktsiy na predpriyatiyakh metallurgii (Corrosion and protection of building structures in metallurgical plants). Stroyizdat, Moscow, p 87–92
16. Verbetskiy VG (1976) Prochnost' i dolgovechnost' betona v vodnoy srede (Concrete strength and durability in water environment). Stroyizdat, Moscow
17. Kozak SI, Nikipanchuk MV, Kotur MG, Grigorash VV (2001) Khimichni osnovi koroziï— konstruktsiynikh materialiv (Chemical foundation of structural materials corrosion). Liga-Press, L'viv

18. Moskvin VM, Ivanov FM, Alexeev SN, Guzeyev EA (1980) Korroziya betona i zkhelezo-betona, metody ikh zaskhckhity (Concrete and reinforced concrete corrosion and protective methods). Stroyizdat, Moscow

19. Fiertak M, Kanka S (1996) Typical cases of sulfate concrete corrosion. In: Mater Conf "KONTRA-96", Warszawa-Zakopane, p 43–45

20. Ivanov FM (1975) Issledovanie cementnykh rastvorov, podvergavshikhsya 68 let deystviyu morskoy vody (Study of cement stones subjected to 68-years-long action of sea water). In: Povysheniye stoykosti betona i zhelezobetona pri vozdeystvii agressivnykh sred (Raising concrete and reinforced concrete resistance to aggressive media). NIIZkhB, Moscow, p 27–30

21. Bulgakova MG (1975) Vliyanie adsorbtsionnoaktivnykh sred na prochnost' i de-formacii betona pri szhatii (Effect of adsorption active media on compressive strength and deformations of concrete). In: Povysheniye stoykosti betona i zhelezobetona pri vozdeystvii agressivnykh sred (Raising concrete and reinforced concrete resistance to aggressive media). NIIZkhB, Moscow, p 38–41

22. Drobyshevskiy BD (1976) The effect of climate factors on deformations and cracking of span structures. Transport Construct 9:14–18

23. Scislewski Z (1999) Ochrona konstrukcji żelbetonowych (Protection of reinforced concrete structures). Arkady, Warsaw

24. Vasil'ev O, Erofeev V, Kartashov V et al. (2004) Protivodeystviye biopovrezhdeniyam na etapakh stroitel'stva, ekspluatatsii i remonta zhilykh i proizvodstvennykh pomeshcheniy (Counteraction against bio damages in stages of construction, exploitation, and repair of residential and industrial rooms). Soft Protector, S.-Peterburg

25. Andreyuk KI, Kozlova IO, Kopteva ZP et al (2005) Mikrobna koroziya pidzemnikh sporud (Microbe corrosion of underground structures). Nauk. Dumka, Kyiv

26. Kopteva ZP, Zanina VV, Purish LM et al (2004) Microflora of impaired reinforced concrete structures in conditions of inhibitor protection. Microbiol J 5:68–75

27. Subbota A, Zakharchenko V, Markevich O et al (2006) Micro fungi impairing structures of buildings. Phys Chem Mech Mater 5:932–936

28. Karyś Y, Kmita A (1999) A case of sulfate and chloride corrosion in coupled concrete pipes. Corros Prot 1:12–14

29. Serednitskiy YA (1994) Zakhist metalokonstruktsiy v gruntakh pidvishchenoy koroziynoy aktivnosti (Metal structure protection in soils with high corrosive activity). In: Ukrains'ke materialoznavstvo (Material science in Ukraine). Phys Mech Inst N.A.S.U., L'viv, p 113–118

30. Serednitskiy Y, Banakhevich Y, Dragilev A (2005) Suchasna protikoroziyna izolyaciya v truboprovidnomu transporti (Modern anticorrosion insulation in pipeline transport). Splain, L'viv

31. Ivanov FM, Vlasov SN (1962) Corrosion protection of reinforced concrete blocks of tunnels. Transport Construct 14:31–34

32. Slobodyan Z, Zvirko O, Kupovich R (2003) Simulation studies of corrosion processes in electrolyte thin interlayer between oil and water. Phys Chem Mech Mater 5:123–124

33. Serednitskiy YA, Teodorovich DA, Kryzhevich LA (1978) Mikro- i biologicheskaya stoykost' poliuretanovykh zashchitnykh pokrytiy (Micro and bio resistance of polyurethane protective coverings). In: Biopovrezhdeniya stroitel'nykh i promyshlennykh materialov (Bio damages of construction and industrial materials). Nauk. Dumka, Kyiv, p 85–86

34. Chandles HT (1979) Corrosion-biofouling relationship of metal in water. Metal Program 115:47–53

35. Serednitskiy YA (2001) Scientific and practical aspects of steel corrosion in presence of sulfate reducing bacteria. Practice Anticor Protect 1:20–30

36. Heinz E, Flemming HC, Sand W (1996) Microbial influenced corrosion of materials: Scientific and engineering aspects. Springer-Verlag, Berlin

37. USSR Standard GOST 10180-90 (1990) Concretes: methods for strength determination using control specimens

38. USSR Standard GOST 28167-91 (1991) Concretes: methods for determining fracture toughness under static loading

39. Rubetskaya TV, Bubnova LS (1971) Methods for concrete damaging depth calculation under corrosive conditions. Concrete Reinforced Concrete 10:18–22
40. Karyś Y, Zubrycki M (1987) Permissible content of cracks in reinforced concrete structures as a function of the amount of reinforcement and corrosion resistance. In: Mater Conf "KONTRA-87", Warsaw, p 87–88
41. Marukha V, Genega BY, Serednitskiy YA (2007) Technology of serviceability restoration using polyurethane injection compositions for concrete and reinforced concrete structures with stress-corrosion cracks. In: Scientific, resource, and technological potential realization efficiency in modern conditions. Proc. 7th Int. Ind. Conf., Lviv, Febr. 2007, p 144–147
42. ACI-89 M12 503-5R (1992) Guide for the selection of polymer adhesives for concrete. ACI Mater J 1–2:90–105
43. Czarniecki L, Skwara Y (1998) Healing cracks in reinforced concrete structures by injection. In: Project execution workshop. XIII All-Poland Conf., Ustroń, p 39–55
44. Golushkova L, Galan' I, Neprila M, Gulay O (2006) Effects of polyester and isocyanate components on viscosity of polyurethane compositions during polymerization. Bull Ternopol Univ 11:31–37
45. Kadurina TI, Omel'chenko SI (1980) Hydrolytic stability and protective properties of ester polyurethanes. Paint Lacquer Mater 3:4–6
46. O'Chaugnessy A, Hoeschale GK (1971) Hydrolytic stability of a new urethane elastomers. Rubber Chem Technol 44:52–61
47. Serednitskiy Y, Banakhevich Y, Dragilεv A (2004) Suchasna protikoroziyna izolyaciya v trubo-providnomu transporti (Modern anticorrosion insulation in pipeline transport). 2nd Part. Splain, L'viv
48. Serednitskiy YA (2001) The effect of polyester block structure and isocyanate components on properties of molded polyurethane elastomers. Composite Polymer Mater 2:45–50
49. Serednitskiy Y (2000) Polyurethane materials for covering main pipelines. Phys Chem Mech Mater 3:84–89
50. Derkach MP, Banakhevich YV, Itkin OF (2003) Experience of gas main pipeline overhaul without stopping gas transport. Oil Gas Ind 4:51–53
51. Czarnecki L, Wysokowski A (2000) Materials for the repair and strengthening of concrete bridge structures. Build Mater 5:40–47
52. Czarnecki L, Jambrozy Z (1996) Material and technology solutions in the repair and protection of concrete structures. Build Mater 8:2–6
53. Czarnecki L (1987) Scientific and technical base of resin injection for objects repair. In: Corrosion protection of modernized buildings and houses. IV Conf. PZITB and Warsaw Center for Progress in Construction, Warszawa–Zakopane
54. Czarnecki L, Scislewski Z (1998) Durability requirements for repaired reinforced concrete structures. Constr Overv 11:4–8
55. Czarnecki L (1998) Materials for the repair and protection of concrete structures. Build Mater 11:8–14
56. Berlin AA, Shutov FA (1978) Penopolimery na osnove reaktsionno-sposobnykh oligomerov (Polymer foams based on reactive oligomers). Stroyizdat, Moscow
57. Bulatov AG (1978) Penopoliuretany v promyshlennosti i stroitel'stve (Polyurethane foams in industry and construction). Stroyizdat, Moscow
58. Dement'ev AG, Tarakanov OG (1983) Struktura i svoystva penoplastov (Structure and properties of foam plastics). Stroyizdat, Moscow
59. Broomfield YP (1996) Corrosion of steel in concrete: Understanding, investigation and repair. Chapman and Hall, London
60. Page CL, Bamforthe PB, Fig JW (1996) Corrosion and reinforcement in concrete construction. Royal Society of Chemistry, Cambridge
61. Marukha VI, Serednitskiy YA, Gnip IP (2008) Characteristics of initial injection composites and solid polyurethanes for renewal of reinforced concrete structures with cracks. Build Mater 71:275–281

62. Maruckha VI, Serednitsky Y (2008) Modeling of destruction processes under extreme loading of systems concrete-polyurethane-concrete. In: Persistence and effectiveness of the repair works. Mater. II Conf., Poznań (Poland), Nov. 2008, p 413–419
63. Marukha VI, Serednitskiy YA, Piddubniy VK, Voloskhin MP (2009) Advanced polyurethane and polyepoxy injection materials for restoration of concrete and reinforced concrete structures. Build Struct 72:465–470
64. Marukha VI, Serednitrskiy YA, Pichugin AT et al (2009) Development and manufacture of pilot equipment, creation of operating shop for production polyol components of polyurethane injection materials. Sci Innov 5:17–24
65. Marukha VI, Gnyp IP (2009) Injection strengthening as a resource-saving technology for repairing concrete structures with cracks. In: Energy and materials saving, environmentally safe technologies. Abstracts of VIII Int Sci. & Tech Conf, Belarus', Grodno, p 81–82
66. Marukha VI, Serednitskiy YA, Voloshin MP (2009) Characteristics of epoxy, epoxy/silicon-organic, and urethane injection compositions and polymers. In: Energy and materials saving, environmentally safe technologies. Abstracts of VIII Int Sci & Tech. Conf., Belarus', Grodno, p 79–80
67. Marukha VI, Serednitsky Y, Voloshin M (2009) Injectable compositions of liquid and polymer inserts to fill cracks and scratches at the renewal of iron-concrete constructions. Plast Chem 6:20–22

Chapter 5
Injection Materials: Technological, Mechanical, and Service Characteristics

Abstract Chapter 5 describes analytical models of and solutions to specific problems concerning strength of deformed bodies with defects filled with injection materials. Obtained solutions are the theoretical basis for service life estimations for structural elements after renewal by injection technologies. For this purpose, a mathematical model of a cracked material healed with injection technologies was developed. The authors analyzed the model in both 2D and 3D formulations. This chapter presents applicability limits of results obtained in two-dimensional approximation. Investigation of crack wedging effects by injection mixtures shows that such effects are significant since they can lead to the growing of the initial crack-like defects under certain conditions.

The extent of filling of defects with an injection material is the important parameter of the technology under consideration. Complete defect filling, in practice, is often hardly attainable, due to various reasons. Therefore, it is important to develop approaches for evaluation of the influence of incomplete defect filling on the effectiveness of damaged structural element renewal.

The problem of injection into a damaged body containing a system of mutually interacting cracks is also considered. The authors consider in detail injection into a system of two cracks. They study the effectiveness of strength restoration for the case of cylindrical structural elements. The solution for transverse compression of a cylindrical element along the planar defect is included.Such specimen configuration is widely used for strength at the testing of brittle materials for strength and fracture toughness. The convenience of such a configuration is in that it requires no special equipment for experiments except for a compression machine. The basic experimental investigations necessary for optimization of injection technologies have been performed using this scheme alone.

5.1 Characterization of Injection Materials

Repair and restorative materials suitable for pressure feeding (injection) into cracks and/or other defects in concrete, reinforced concrete, or brick structures include three principal classes: cement-mineral, cement-polymer, and polymer.

The basic preparation method for paste-like cement-mineral or cement-polymer materials is mixing the cement, mineral fillers, and polymer modifiers with water. Dilution of thick paste-like cement-based compositions with water allows for an

V. V. Panasyuk et al., *Injection Technologies for the Repair of Damaged* 97
Concrete Structures, DOI 10.1007/978-94-007-7908-2_5,
© Springer Science+Business Media Dordrecht 2014

increase in the ability to inject, if needed. The list of various polymer injection materials includes single-component and dual-component polymer compositions based on the components capable of polycondensation or polymerization at $20 \pm 20\,°C$.

Implementation of technological procedures for crack or damage filling with various injection materials in concrete matrices requires initial consideration of all the service limitations of these materials in relation to permissible crack opening and unfilled crack tip area length, in particular:

- Cement mortar (ZS) as filling material requires a crack opening wider than 0.8 mm and crack tip areas as large as 0.2 mm in length remain unfilled;
- Cement suspension filler (ZL) requires a crack opening wider than 0.2 mm and areas greater than 0.05 mm in length remain unfilled;
- The polyurethane or polyepoxy resin-based fillers (PU or EP, respectively) require a crack opening wider than 0.1 mm and no unfilled areas remain.

5.2   Cement-based Injection Materials

Cement-based injection materials with mineral fillers and/or water-soluble polymer modifiers find wide application in the strengthening and repairing of concrete structures. The technical literature refers to such materials as aqueous cement suspensions or concrete mortars. The primary advantage of such materials is close matching between the mechanical characteristics of solid composites, which form from them after curing, and those of the parent concrete matrix.

Cement or cement-polymer suspensions based on fine grain components, such as the compositions ZL made by MC-Bauchemie Müller GmbH (Germany), are applicable at crack openings over 0.2 mm. At crack openings over 0.8 mm, ordinary cement mortars based on coarser mineral fillers are permissible, for example, compositions ZS made by the same firm.

Injection materials of a cement or cement-polymer basis in the form of aqueous suspensions or concrete mortars (Table 5.1) are suitable for non-pressure repair of dry or wet cracks. The aqueous cement or cement-polymer suspensions of the type ZL or ZS are injectable into cracks or defects with opening widths over, respectively, 0.2 mm or 0.8 mm. The particle size of the cement base and mineral fillers also determines the selection of feeding pumps that can operate with fine or coarse injection materials (see Fig. 3.5 and Table 3.2).

Mineral filler content in concrete mortars is a very important factor for repair or restorative injection. Most effective for feeding into cracks and damages of concrete structures are cement or cement-polymer mortars containing 'floating' fillers. The equipment for feeding the aqueous concrete mortars (suspensions) (Table 3.2) into cracks and defects allows for the use of insoluble fillers with a grain size from 0.3 to 3.0 mm.

Technical requirements for cement-based injection materials establish the following parameters:

Table 5.1 Basic parameters of cement-polymer injection materials produced by MC-Bauchemie Müller GmbH (Germany) and Sika Services AG (Switzerland)

Parameter name and dimensions	Centricrete (MC-Bauchemie)			Sika®Injecto Cem-190
	UF	FB	MV	
Density, g/cm^3	1.7	1.7	1.79	1.7
Time of efflux (flowability), s	50	50	60	100
Component ratio, weight parts	10: 5: 0.9	20: 12: 0.4	20: 9: 0.4	5: 3.2
Application time, min	57	30	60	120
Minimum application temperature, °C	+8	+5	+5	+5
Compressive strength, MPa:	11	22	17	40
2 days	15	36	23	44
7 days	31	44		47
28 days				
Tensile strength, MPa:	4	3	2.6	
2 days	7	3.7	4.5	
7 days	7.2	4		
28 days				
Change in volume, %	2	2	0	0.25
Permissible crack opening width, mm	≥ 0.25	≥ 0.6	≥ 0.6	≥ 0.2

- Grain size of initial cement base and mineral fillers;
- Viscosity retention time and absence of premature paste solidification;
- Permissible settlement of solid mineral particles in aqueous concrete suspensions;
- Corrosion resistance of the cement base, fillers, and solid concrete

The injection cement or cement-polymer suspensions (mortars) fail to fill in cracks or damages with an opening width below 0.2 mm effectively because of insoluble fine and/or coarse particles. On the other hand, such paste-like concrete mortars can cause difficulties when being fed into cracks or damages under pressure, even that as low as 20 to 30 degrees of atmosphere. (AU: I'm not sure I've gotten this right, but 'a few tens' is definitely not an expression with which I'm familiar.) At the same time, fluent polyurethane or polyepoxy compositions allow pumping at pressures of 10–150 atm, ensuring penetration into microcracks and capillaries of concrete matrices.

In addition, heterogeneity of aqueous concrete mortars bearing mineral particles 0.3–3.0 mm in size creates the risk of clogging the flow passages during injection. This phenomenon induces essential technological obstacles for the stable filling of cracks and damages in concrete structures with cement suspensions (Fig. 5.1).

5.3 Polymer Injection Materials

Polymer materials have essential advantages over aqueous cement or cement-polymer suspensions (mortars) in strength and serviceability restoration of internally cracked concrete or reinforced concrete structures [1]–[5]. Therefore, utilization of

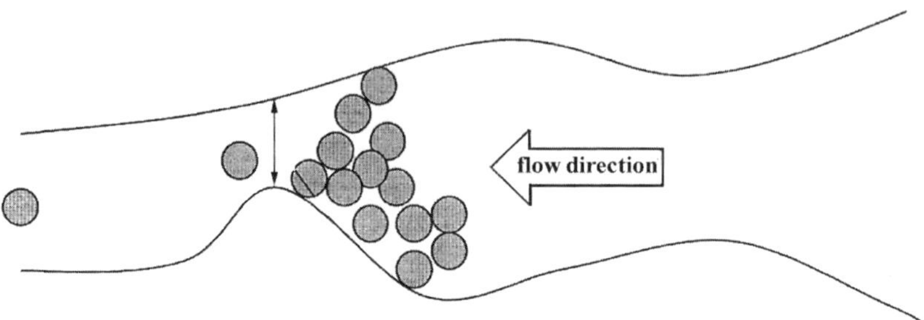

Fig. 5.1 Illustration of clogging in crack-like defects due to feeding concrete mortar with solid particles

such materials is one of most effective technological approaches for the protection of impaired objects from further failure, particularly stress corrosion fracture. Moreover, injection of pressurized fluent polymer compositions into pre-determined zones indicates the use of diagnostic procedures and enables the restoration of the damaged sections operatively, i.e., without stopping use of the concrete structures.

Polymer injection material embraces a wide scope of polymer compositions capable of retaining fluidity after preparation until pressurization and feeding (injection) into inner defects and solidification in the repaired concrete matrix due to polycondensation or polymerization (only polyacryl materials). The dual-component fluent viscous compositions of polyurethane, polyepoxy, polyester, polyacrylic, or silicon-organic base are most commonly used. The common name of the above materials is 'polymer injection materials' or simply 'injection materials'.

Polymer injection materials and methods of injection compose one of the most widely used in the Ukraine processes for repair and restoration works in concrete, reinforced concrete, and brick structures and buildings. Such materials have obtained approval in the regulations of the State Committee for Construction of Ukraine regarding the use and repair of buildings, structures, and engineering mains [6], [7], as well as branch regulations regarding safe and reliable use of industrial buildings and structures and rules of residential houses and maintenance of adjacent territories [7], [8].

The foreign technical literature [3]–[5] and datasheets of foreign firms represent a wide range of injection materials also based on fluent polymer compositions. Note that the foreign manufacturers of injection materials and suppliers of repair or restorative service pay the most attention to polyurethane compositions.

Classification of polymer injection materials relies on the basic initial monomer or oligomer compound type: polyurethane, polyepoxy, polyacrylic or silicon-organic. Primary polymer selection criterion (Table 5.2) is the status of the cracks or damages in the concrete matrix of the impaired object:

- Dry or slightly wetted damage, e.g., due to moisture condensation, or
- Heavily wetted due to water leakage or seepage (hydraulic works, water mains, sewage collectors, etc.) 4.4. General specifications and advantages of polymer injection materials

Table 5.2 Crack wetting influence on injection material characteristics

Injection materials for dry cracks	Injection materials for wet cracks
Viscosity at 20 °C is below 120 mPa·s	Viscosity at 20 °C is over 150 mPa·s
Viability (gelation time) is 70–120 minutes	Viability (gelation time) is 5–20 minutes
Volume increase is by 1.1–2.0 times	Reaction delay time after contacting with water is 5–50s
–	Volume increase is by 8–40 times

General requirements for polymer injection materials include the following:

- Preparation by mixing of two liquid low-viscous components at normal conditions, the components being monomers or oligomers of the type 'base' (the component A) and 'hardener' (the component B);
- Formation of homogeneous fluent compositions after mixing at normal temperatures that are suitable for pressurization and feeding into cracks or damages in concrete structures;
- Wetting surfaces of concrete matrices and penetration into large and small cracks, capillaries, voids, etc.;
- Impregnation of surface layers of the concrete matrix in cracks and damages with polymer compositions and/or individual components;
- Complete or maximum possible filling of cracks and damages in concrete structures;
- Capability of structurization at normal conditions (20 ± 2 °C) with formation of solid polymer inserts;
- Presence of functional groups in the polymer structure capable of chemical interaction with reactive groups of cement stone;
- Presence of functional groups in the polymer structure capable of chemical interaction with water molecules with formation of high-molecular compounds and locking water paths through cracks or damages

Technical requirements for solidified polymer injection materials include the following:

- Insignificant (below 2.0 vol. %) values of volume change (expansion or shrinkage) during hardening;
- Maximum possible volume filling of cracks and damages;
- Strong adhesive (chemical and physical) bonding to constituents of concrete matrices;
- Formation of a single composite structure of concrete-polymer-concrete during hardening;
- Greatest possible prevention of access of water and contained corrosive compounds and microorganisms to concrete surfaces in healed cracks and damages;
- Parameters of strength and elasticity higher than those of the concrete;
- Restoration of strength and serviceability of damaged concrete structures;
- Long-term conservation of operational characteristics

The advantages of polyurethane injection compositions and solid polyurethanes over other similar materials have found confirmation in the results of field trials of restoration of strength and serviceability of damaged concrete structures [1]–[3], [9]. It was established during the restorative works in the concrete structures of Ukraine that the most effective polymer compositions were those capable of, on the one hand, most completely filling cracks or defects, and on another hand, hardening fastest. As a rule, polyepoxy, polyacrylic and silicon-organic injection materials had demonstrated less efficiency than the polyurethane materials [10]–[13]. Therefore, implementation of polyurethane injection materials in the practice of repairing damaged concrete structures is well grounded [12], [13].

First, the dual-component polyurethane compositions have higher viability as compared with silicon-organic or polyepoxy materials that finds expression in longer fluidity retention before initial polymerization (stiffening or filament formation) [14], [15]. The intentional selection of the polyol and/or isocyanate components permits for regulation of rates of the polyurethane forming reactions. The optimal viability ensures easy preparation and practical application of the fluent polyurethane compositions during injection under pressure into defects of concrete structures.

Second, the relatively high viability after mixing base (the component A) and hardener (the component B) combines well in the polyurethane reactive compositions with the beginning of solidification (stiffening), which is roughly less than 10 min. These process parameters are similarly subject to regulation by changing type and proportions of the initial polyol and isocyanate components [15]. Besides, polyurethane compositions are notable for high molding ability, low shrinkage during solidification, and absence of complementary internal stresses in contact zones near structural material surfaces due to high elasticity [16].

Third, an essential technological advantage as compared with polyepoxy or other polymer compositions exists in the ability of polyurethane materials, in particular, isocyanate components of these materials (component B), to structurize with the participation of water molecules. The technical importance of this parameter grows considerably in the cases of the serviceability renewal of concrete sewage collectors, hydraulic works, and seaport constructions. Note that the smallest water drops on the smooth surface of such concrete structures are virtually unavoidable, not to mention water in pits, cracks, and other defects. The references [17]–[19] contain detailed information on this aspect.

Chemical interaction of pre-polymer's end isocyanate groups with water molecules in the specifically composed polyurethane compositions causes foaming. The reaction proceeds with the evolution of great volumes of carbon dioxide gas and respective foaming of the polyurethane mass in the course of polymerization. The increase in polyurethane volume as compared with the volume of the initial composition can be as high as 10–40 times. If such interaction with water occurs in a closed space, for example, a crack or void in the concrete, the resulting foamed polymer with closed cells blocks any paths of water leakage or seepage.

At the same time, compact (non-foamed) polyurethane elastomers are flexible and, hence, possess high deformation characteristics that ensure, in particular, the service of reliable coverings and injected polymers at high mechanical static or cyclic loads (compression, tension, bending, shear and/or friction wear during earth moving, vibrations, etc.) [20], [21].

Another confirmation of the efficiency of polyurethane compositions in healing cracks and damages in concrete matrices is obvious from the fact of prolonged service life of concrete/polyurethane elastomer/concrete junctions under high loads. For example, the tensile strain of a concrete structure bound with a polyurethane binder can be as high as 5 %.

Next, solid injected polyurethanes formed due to polycondensation in cracks and defects inside the concrete mass must have a high water-resistant property. This property makes the aqueous corrosive media protection of damages in restored concrete structures more reliable and prolonged. Many domestic and foreign researchers have substantiated the suitability of polymer injection materials for service in such conditions by means of testing water resistance of polyurethane elastomers depending on qualitative composition and structure [22]–[25].

5.4 Advantages of Polyurethane Compositions and Elastomers Over Other Polymer Injection Materials

Domestic fluent cold hardened polyurethane compositions [16] comply with the above technical requirements for injection materials. Such compositions have the component A (base) to component B (hardener) mass ratio of 100:11–18. The mixing temperature is 20 °C or 50–60 °C; fluidity duration at 20 °C or 50–60 °C is, respectively, 2.5–4.0 hours or 1.5–2.0 hours; viscosity determined as an efflux time in the viscometer VZ-4 is 50 min at 50 °C; solidification time in concrete forms at 20 °C is 1.0–1.5 days, while at 65 °C it is 5.0–8.0 hours.

The solid polyurethane elastomers from the above compositions have the following characteristics: cured density 0.956 g/cm^3; shrinkage during hardening 7 %; Shore hardness A 50–55 °; ultimate tensile strength 3.5 MPa; tensile strain 600–700 %, and water uptake (wt. %) at 20 °C 0.09–0.13 per 24 hours or 0.50–0.55 per 60 days while at 100 °C 2.3 per 35 days.

Injection composition Webak 1403 produced by WEBAC® Chemie GmbH (Germany) complies with the above requirements for polyurethane materials as well [26]. The Webak 1403 is a low-viscosity dual-component polyurethane resin containing no organic solvents. Its characteristics are as follows: the components A to B volume ratio 1:1; component A density 1.01 g/cm^3 and component B density 1.15 g/cm^{3at} +20 °C; viscosity of the composition 80 MPa·s at +20 °C; application time 1.5 hours; application temperature over +5 °C; Shore hardness A 35 °.

Determination of Fluidity Retention Time for Polyurethane Injection Compositions in Laboratory and Field Conditions

Low viscosity retention after mixing of the reactive components and its slow growth during injection is the principal technological parameter of polyurethane compositions. The fluidity retention becomes possible primarily due to the absence of

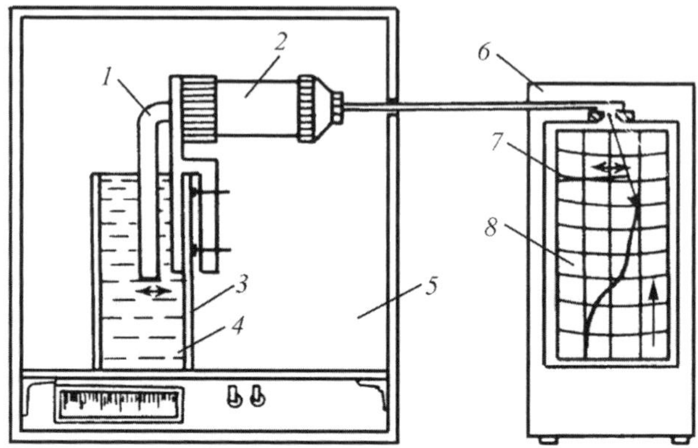

Fig. 5.2 Layout of the device for viscosity change control during polymerization of polyurethane compositions. Shown in the diagram are: vibration probe (*1*); vibration generator (*2*); composition sample containing reservoir (*3*); composition sample (*4*); thermostat (*5*); registration units (*6–8*)

premature structurization with formation of a so-called gel skeleton (i.e. stiffening or filament formation). The above phenomenon makes the respective polyurethane oligomers unsuitable for injection because of high viscosity.

The simplest method for determining viscosity of fluent chemical products under laboratory conditions consists of measurement of the efflux time for a certain liquid volume flowing through a calibrated orifice between two marks. In this case, the viscosity value can be calculated using the formula:

$$\eta = K\rho\tau$$

where,

η is the dynamic viscosity of the liquid
K is the instrument constant
ρ is density of the liquid, and
τ is the efflux time.

Depending on the measurement instrument (viscometer) type, the viscosity value (η) for the fluent polymer compositions is representable in time units (minutes or seconds). However, the above-presented viscosimetric method is applicable only for the polymer compositions that either do not structurize at all or have a long stiffening time (e.g., about 50 min, as with the cold hardened polyurethane compositions described in this section).

Conversely, highly reactive dual-component polyurethane compositions can structurize with formation of a gel skeleton (stiffening) over a few minutes that can cause clogging and failure of the viscometer. For this reason, the indirect methods are most common for measuring the viscosity changes during structurization under laboratory conditions. For instance, vibrational oscillations of the PKV-2 device sensor yield time dependence of the viscosity growth during polycondensation of the polyurethane compositions at $20 \pm 2\,^{\circ}\mathrm{C}$ (Fig. 5.2). Such kinetic curves, besides the primary parameter and the fluidity retention time, also illustrate the structurization behavior of polyurethane systems with different base components (Figs. 5.3, 5.4) [14].

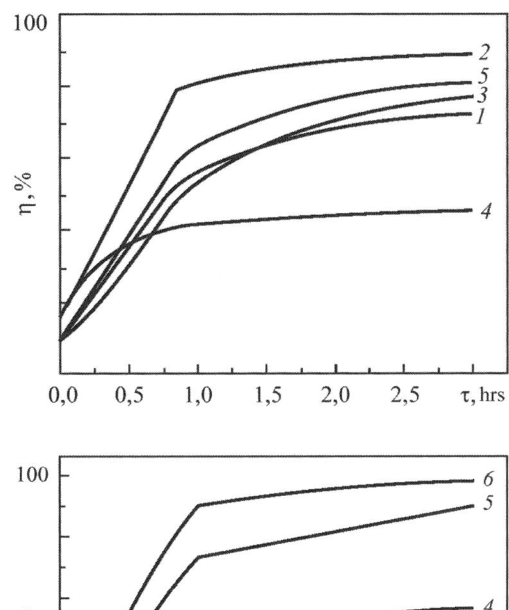

Fig. 5.3 Viscosity change curves for polyurethane compositions based on polyester P-2200 with various isocyanates: mix of 2,4 and 2,6-toluylene diisocyanates T-65–35 (*1*); polyisocyanate PIC (*2*); 2,4-toluylene diisocyanate 102-T (*3*); 4,4'-diphenylmethane diisocyanate MDI-T (*4*)

Fig. 5.4 Viscosity change curves for polyurethane compositions based on polyesters (PE) or PIC: PE-2200 (*1*); EDA-50 (*2*); PE-2200 + PDA-800 (1:1) (*3*); PE-2200 + PEG-3 (1:1) (*4*); P-2200 (*5*); PDA-800 (*6*)

The experimental data (Figs. 5.3 and 5.4) lead to the conclusion that the viscosity stabilization in the time interval of 0.7–3.0 hours after the curve kink corresponds to formation of a gel skeleton (stiffening) in the polyurethane compositions. Samples of fluent polyurethane compositions under laboratory conditions reach Shore hardness A 25–30° due to polymerization after the same time interval [15], [16]. The similar technological test or trial for determination of the primary parameter, the fluidity retention time, is suitable for field measurements as well. The above presented data (Fig. 5.3) also characterize the reactive capability of polyesters and isocyanates at the section of fast viscosity growth before the gel point (kink of the curve).

The kinetic curves showing the fluidity retention time until primary structurization (gel or stiffening point) of polyurethane compositions are basic characteristics when selecting the injection material formulation and developing process for feeding of the pressurized compositions into cracks or damages of concrete structures. In-lab or field tests or trials are supplementary to the above measurements. Such express tests also allow for estimation of the fluidity retention time of polyurethane compositions.

In practice, the fluidity retention time is determined in relation to the formation of a gel skeleton through visual observation and tests of the stiffening or filament

Table 5.3 The influence of polyurethane formulation on the deformational properties of elastomers [10]

Sample No.	Polyester	Isocyanate	Compressive strain, %	Residual strain, %
1	P-2200	PIC	12.20	0.13
2	P-2200	MDI-T	5.99	0.05
3	P-2200 + L-294 (1:1)	PIC	4.92	0.02
4	P-2200 + PDA-800 (2:1)	PIC	8.19	0.32
5	P-2200 + PDA-800 (1:1)	PIC	7.42	0.87
6	PDA-800	PIC	4.84	0.09
7	PE-2200 + PEG-3 (1:1)	102-T	4.56	0.13

formation condition using a glass stick. This method establishes the closing stage of process, structurization of fluent injection materials and formation of primary polyurethane oligomers. The fluidity retention time for polyepoxy, polyacrylic, or silicon-organic polymer materials is determined in a similar way.

Physical and Mechanical Characteristics of Solid Injected Polyurethanes

The method of technological test or trial is useful for establishing not only fluidity retention time, but also the mechanical, chemical, and service characteristics of solid polyurethane elastomers. Such characteristics are especially useful for predicting the operational efficiency of polyurethane inserts in concrete structures and correction, if needed, of the formulation, structure, and properties.

Let us consider, for example, the influence of formulation of the complex polyesters and polyisocyanates containing polyurethane compositions on the deformational physical and mechanical properties of polyurethanes (Table 5.3). We can see that the linear polyester PDA-800 with minimum molecular mass forms with 2.4-toluylene diisocyanate 102-T a strong weakly elastic polyurethane revealing Shore hardness A 45°. However, the same PDA-800 with the polyisocyanate PIC forms the polyurethane with a three-dimensional structure revealing Shore hardness A 55° and elevated tensile strength, ultimate strain, and elasticity due to the branching structure of the polymer isocyanate (Table 5.4).

We can note from Table 5.3 that the synthesis of linear polyester PE-2200 in reaction with isocyanate 102-T, on the contrary, provides a decrease in polyurethane hardness down to Shore A 34°, tensile strength σ_b down to 0.90 MPa, and residual strain down to 2.03 %. Similar softening down to Shore hardness A 41° is also notable in the system PE-2200 + PIC, although here σ_b grows up to 2.15 MPa [15].

The additional structuring with oxypropylated ethylenediamine (Lapromol-294) in the system PE-2200 + PIC increases Shore hardness A up to 48° but decreases strength σ_b down to 1.27 MPa.

Table 5.4 The influence of complex polyesters and isocyanates structure on the properties of polyurethane elastomers [15]

No.	Polyester	Isocyanate	Hardness, ° Shore A	Ultimate strength, MPa	Ultimate tensile strain, %	Residual strain, %	Rebound elasticity, %
1	PDA-800	102-T	45	1.82	233	5.13	21
2	P-2200	102-T	35	1.82	350	0.85	38
3	P-V	102-T	36	1.53	347	1.09	35
4	PE-2200	102-T	34	0.90	169	2.03	31
5	EDA-50	102-T	38	2.41	282	2.87	25
6	PDA-800	PIC	55	2.52	453	5.31	24
7	P-2200	PIC	32	1.78	287	0.90	28
8	PE-2200	PIC	41	2.15	272	6.70	26
9	PE-2200 PDA-800 (1:1)	PIC	35	0.89	270	6.12	23
10	PE-2200 EDA-50 (1:1)	PIC	36	1.27	277	3.12	22
11	PE-2200*	PIC	48	1.36	260	6.81	24
12	EDA-50	PIC	40	2.53	279	3.03	26

* Polyurethane elastomer cross-linked by Lapromol-294

The linear polyester EDA-50, based on ethylene glycol (molecular mass 1850) in the systems with 102-T and PIC, strengthens the materials up to 2.41 MPa and 2.53 MPa, respectively. However, growth in molecular mass and decrease in the polar urethane group concentration reduce Shore hardness A down to 38–40° in comparison with the Shore hardness A 45–55° for PDA-800 (Table 5.4). At the same time, the effect of the isocyanate component on the properties of polyurethanes based on complex polyester P-2200 is negligible (Table 5.5) [15].

The values of residual strain are very close for all polyurethanes and vary within 0.02–0.13 % (Table 5.3). Only the systems P-2200 + PDA-80 with PIC are essentially worse in this relation (0.32 % and 0.87 %). In other words, the branched materials possess better deformational properties due to polyester (P-2200, PEG-3) and polyisocyanate (PIC) blocks.

Polyurethane-injected junction inserts have to be waterproof. To ensure such a property, the waterproof polyester components must present in the formulation. However, polyurethanes on complex polyesters have a hydrolytic stability insufficient for long-term service. The water intake data of Table 5.6 as well as other characteristics of polyurethanes based on simple or complex polyesters and polydiene diols confirm this feature [16], [22]–[25].

Table 5.5 The influence of the isocyanate component on the properties of polyurethane elastomers based on greatly branched polyester P-2200 [15]

Parameter name	2,4-toluylene diisocyanate 102-T	Mix 65/35 of 2,4- and 2,6-toluylene diisocyanates T-65–35	Equimolecular mix of T-65–35 and polyisocyanate PIC	Polyisocyanate PIC	4,4'-diphenyl-methane diisocyanate MDI-T
Hardness, ° Shore A	35	28	40	32	31
Ultimate tensile strength, MPa	1.82	2.10	1.72	1.78	2.34
Tensile strain, %	350	288	226	287	237
Residual strain, %	0.85	0.99	0.85	0.90	0.70
Rebound elasticity, %	38	30	24	28	37

5.5 Characteristics of Fluent Polyurethane Compositions and Solid Elastomers

Polyurethane foams (PUF) are one of most common and technologically convenient foam plastics in both production and utilization. Therefore, world production volumes of polyurethane foams are twice as high as the volumes of all other foam plastics [27]–[29]. The chemism of polyurethane foaming is as follows. Interaction between excess diisocyanates or polyisocyanates (the component B) and hydroxyl containing polyols (the component B) during preparation of the injection compositions results in the formation of polyurethane fragments, so-called pre-polymers. On arriving at the destined cracks in the concrete matrices, end groups of the pre-polymers react with water. Evolving in these reactions, carbon dioxide gas uniformly foams into a still fluent polyurethane mass. Low-boiling liquids, e.g., chlorofluorocarbons (freons), can provide an additional foaming if introduced into polyols (the component A) during polyurethane formation. Polyurethane formation and foaming reaction rates can be regulated using catalysts [29], [30].

Regulation of the extent of polymer macromolecule branching allows for the production of rigid or elastic polyurethane foams. Therefore, mass polyurethane products include simple and complex polyesters, catalysts, and foaming agents in formulations of the component A and diisocyanates of the diphenylmethane series or polyisocyanates in formulations of the component B [25]–[30]. Solid polyurethane foams possess chemical, environmental, and microbiological stability at a wide range of operational temperatures; high strength; resistance against boiling water, and negligible water and vapor permeability. These features enable the widest applications of polyurethane foams. Modification of polyurethanes with the silicon-organic polyols can further enhance their stability.

Table 5.6 The influence of the polyester block on the properties of polyurethane elastomers based on 2.4-toluylene diisocyanate

Parameter name	Polydiethylene glycol adipinate P-2200	Polydiethylene glycol adipinate P-2200, modifier methylmetacrylate	Polyoxypropylene glycol (Laprol 2002)	Copolymer of tetrahydrofuran with propylene oxide (Laprol 1102-4-80) (Fergana chemical plant)	Copolymer of tetrahydrofuran with propylene oxide (Laprol 1102-4-80) (Vladimir chemical plant)	Polydiene diol PDI-1K
Hardness, ° Shore A	48–50	64–66	64–66	68–70	73–74	61–63
Ultimate tensile strength, MPa	2.3	6.7	5.2	13.0	21.5	3.5
Ultimate tensile strain, %	510	190	220	734	737	669
Relative water uptake per 72 hours, %	2.42	2.00	2.37	1.61	2.41	0.13

Table 5.7 Physical and chemical characteristics of polyurethane compositions according to specifications TU U V.2.7-24.1-13803953-017-2008 before mixing

Parameter name	Component A			Component B "Technonat"	Testing method acc. specification
	Technopol 5012	Technopol 2970	Technopol 3301		
Appearance	Layered liquid from yellow to brown color			Transparent liquid from white to brown color	5.1.1
Hydroxyl number, mg K OH/g	From 110 to 150			–	5.1.2
Acid number, mg K OH/g	From 11 to 16			–	5.1.3
Dynamic viscosity, at 21 °C, MPa	From 500 to 700	From 450 to 650	From 250 to 450	From 150 to 290	5.1.4
Max. water content, wt. %	0.5			0.5	5.1.5

Table 5.8 Physical and chemical characteristics of polyurethane compositions according to specifications TU U V.2.7-24.1-13803953-017-2008 after mixing but before solidification

Parameter name	Rated value depending on the component A Technopol:			Testing method acc. specification
	Technopol 5012	Technopol 2970	Technopol 3301	
Start time, s	From 5 to 7	From 23 to 33	From 14 to 20	5.3.1
Gel formation time, s	From 12 to 14	From 120 to 170	From 65 to 85	5.3.2
Growth time, s	From 12 to 16	From 170 to 220	From 90 to 120	5.3.3
Viability of the polyurethane system, s	From 7 to 9	From 110 to 120	From 55 to 65	5.3.4

Polyurethane Foam Applications for Injection Renewal of Concrete Structures Damaged by Leaking or Seeping of Water

Polyurethane foamed composite materials formed in cracks or damages of hydraulic concrete structures can block and completely cease water seepage and even leakage under pressure. Implementation of this process in practice requires the polyurethane compositions to react readily with water, i.e., contain pre-polymers with end isocyanate groups. According to specifications TU U V.2.7-24.1-13803953-017-2008 "Polyurethane system TECHNO-PUR" of SPRC "Techno-Resource" N.A.S. Ukraine, foaming dual-component polyurethane compositions must comply with the following requirements (Tables 5.7–5.9).

Table 5.9 Physical and chemical characteristics of polyurethanes according to specifications TU U V.2.7-24.1-13803953-017-2008 after solidification

Parameter name	Rated value depending on the component A Technopol:			Testing method acc. specification
	Technopol 5012	Technopol 2970	Technopol 3301	
Density, kg/m^3	Over 55	From 26 to 32	From 28 to 33	5.4.1
Material structure and appearance	Closed cellular mass from yellow to brown color			5.4.2
Min. specific volume of closed air-filled cells, %	90			5.4.3
Mechanical compressive strength, kPa	From 200 to 300			5.4.4
Heat conductivity coefficient, W/(m·K)	0.03			5.4.5
Max. thermal stability of insulating layer, °C	Plus 130			5.4.6
Max. water uptake, %	3	4	2	5.4.7
Adhesion to concrete surface, MPa	From 1.5 to 2.0			5.4.7
Min. cold resistance, °C	Minus 40			5.4.6

Injection composition Webak 157 produced by WEBAC® Chemie GmbH (Germany) complies with requirements similar to the polyurethane materials [26]. Namely, it is a low-viscous dual-component polyurethane resin containing no organic solvents. It has an A to B ratio of 5:1; densities of 1.1 g/cm^3 (A) and 1.0 g/cm^3 (B) at $+20$°C; viscosity of reactive composition 320 MPa·s at $+20$°C; viability of reactive composition 2 hours at $+20$°C; application temperature over $+5$°C; foaming start time after contact with water 20 s, and foaming finish time 130 s.

Volume increase of the polyurethane foamed composite materials formed in a concrete mass reaches 15–30 times. Minimal hardening time is less than five minutes [26], [27], [29]. The hardened PUF forms adhesion bonds with the constituents of the concrete matrices in the crack tips and between crack edges just like the non-foamed polyurethanes. As a result, the damaged concrete structure becomes stronger and simultaneously all paths of water penetration through damages in the concrete become blocked.

However, adhesive bonds of polyurethane foams with concrete are weaker than the bonds of compact polyurethane due to 15–30-fold volume expansion. Moreover, despite the closed cellular structure of polyurethane foams, corrosive media dangerous to concrete can penetrate through individual defects of the cellular structure.

For these reasons, with the aim of densifying the above-mentioned sections of the structure, the additional non-foaming polyurethane compositions fed under pressure through the drilled channels in 10–30 min after forming the injected polyurethane foam filling are useful. In such a manner, the solid polyurethane elastomer arises in the defect zones of polyurethane foam constituents in the composite system of concrete/polyurethane foam/concrete. The solid polyurethane elastomer enhances both the strength and waterproofing properties of the foamed repair and restorative material, as well as its composite with the concrete.

5.6   Polyepoxy, Polyacrylic, and Silicon-organic Injection Materials

Well-known European firms such as WEBAC® Chemie GmbH and MC-Bauchemie Müller GmbH (Germany), PHU TECH-KAN Adam Wojciechowski (Poland) and others, use various dual-component injection materials on the polyepoxy, polyacrylic, and silicon-organic bases to restore strength and serviceability of the cracked concrete structures. Nevertheless, application volumes of such materials are much less than the volumes of polyurethane analogues. Explanation of this superiority lies primarily in the specific features of repaired structures, diversity of defects, processing advantages of fluent polyurethane compositions, and the high service parameters of solid polyurethanes and polyurethane foams.

The basic specifications of the polyepoxy, polyacrylic, and silicon-organic injection compositions and solid polymers based on these compositions are as presented below. Note that the technology of injecting both groups of materials into cracks or damages of a concrete matrix is virtually the same.

Injection Resin Webak 4110 of WEBAC® Chemie GmbH (Germany) [26]:

The material has the form of a viscous dual-component polyepoxy resin without organic solvents. Components volume ratio A:B = 2:1. Density of component A is $1.13 \, g/cm^3$, component B $0.94 \, g/cm^3$ at $+20\,°C$. Viscosity of reactive composition is 280 MPa·s at $+20\,°C$; viability of reactive composition is 100 min at $+20\,°C$; application temperature is $+8\,°C$ or higher; hardening time is 24 hours at $+20\,°C$.

Injection Polyacrylic Gel Webak 250 of WEBAC® Chemie GmbH (Germany) [26]:

The material has the form of a low-viscous triple-component water-soluble polyacrylic resin. Components volume ratio (A1 + A2):B = 1:1. Density of component mix (A1 + A2) is $1.05 \, g/cm^3$, component B $1.0 \, g/cm^3$ at $+20\,°C$. Viscosity of mix (A1 + A2) is 5.0 MPa·s, component B 1.0 MPa·s at $+20\,°C$. Viscosity of reactive composition is 2.0 MPa·s at $+20\,°C$; viability of reactive composition is 13 min at $+20\,°C$; application temperature is $+5\,°C$ or higher.

Injection Composition (Resin) MC-Inject GL-95 TX of MC-Bauchemie Müller GmbH (Germany) [27]:

The material has the form of a low-viscous water-soluble acrylic resin without organic solvents enhanced with polymer additives. Components mass ratio A:B = 99–96: 1–4. Viability at $+20\,°C$ as a function of the component B content is 20 s at 4 wt.%; 37 s at 2 wt.%, and 59 s at 1 wt.%.

Hardening and Strengthening Injection Material Webak 2061 of WEBAC® Chemie GmbH (Germany) [26]:

The material has the form of a low-viscous dual-component modified silicon-organic resin without organic solvents. Components volume ratio A:B = 10:1. Density of component A is 1.24 g/cm^3, component B 1.15 g/cm^3 at + 20°C. Viscosity of component A is 10.0 MPa·s, component B 3.0 MPa·s at + 20°C. Application temperature is + 10 °C or higher, viability of reactive composition is 20 min at + 20°C.

References

1. Marukha VI, Genega BY, Serednitskiy YA (2006) Efficiency of polyurethane injection materials application for serviceability restoration in concrete or reinforced concrete structures with stress corrosion cracks. In: Diagnostika, dovgovechnost' ta rekonstruktsiya mostiv i budivel'nykh konstruktsiy (Diagnostics, durability, and reconstruction of bridges and concrete structure). No. 8. Kamenyar, Lviv, p 84–90
2. Marukha A, Genega B, Serednitskiy Y, Zaplatins'kiy M (2006) Concrete structure protection against stress corrosion using polyurethane injection compositions. Phys Chem Mech Mater 5:834–840
3. Czarnecki L, Emmons PH (2002) Naprava i ochrona konstrukcji betonowych (Repair and protection of concrete structures). Polski Cement: Krakov
4. Jankowski P (1993) Sealing of reinforced concrete structures using chemical injection. Engin Construct 3:101–104
5. Podolski B, Suwalski J, Wydra W (2000) The specifics of repair of old reinforced concrete building structures. Corr protect 1:67–69
6. Esipenko AD (2006) System formation principles for houses, structures, and engineering mains maintenance and repair. Construction in Ukraine 1:36–38
7. Regulations regarding expection, passportization, safe and reliable exploitation of industrial buildings and structures (1997). R & D Institute for Industrial Construction: Kyiv
8. Esipenko AD, Konopko NP, Starodubneva LV (2005) Pravila utrimannya zhilykh budinkov ta prybudynkovykh territoriy (Rules for maintenance of residential houses and adjoining areas). R & D Institute for Industrial Construction: Kyiv
9. Czarnecki L, Jambrozy Z (1996) Material and technology solutions in the repair and protection of concrete structures. Build Mater 8:2–6
10. Marukha VI, Serednitskiy YA, Gnip IP (2008) Characteristics of initial injection composites and solid polyurethanes for renewal of reinforced concrete structures with cracks. Build Mater 71:275–281
11. Marukha VI, Serednitskiy YA, Pi ddubniy VK, Voloskhin MP (2009) Advanced polyurethane and polyepoxy injection materials for restoration of concrete and reinforced concrete structures. Build Struct 72:465–470
12. Marukha VI, Genega BY, Serednitskiy YA (2007) Technology of serviceability restoration using polyurethane injection compositions for concrete and reinforced concrete structures with stress-corrosion cracks. In: Scientific, resource, and technological potential realization efficiency in modern conditions. Proc. 7th Int. Ind. Conf., Lviv, Febr. 2007, p 144–147
13. Marukha VI, Serednitskiy YA, Voloshin MP (2009) Mechanism of chemical and physical interaction in injection formed repairing systems concrete-polyurethane-concrete. Engin Mech 25:228–232
14. Golushkova L, Galan' I, Neprila M, Gulay O (2006) Effects of polyester and isocyanate components on viscosity of polyurethane compositions during polymerization. Bull Ternopol Univ 11:31–37

15. Serednitskiy YA (2001) The effect of polyester block structure and isocyanate components on properties of molded polyurethane elastomers. Composite Polymer Mater 2:45–50
16. Serednitskiy YA, Golushkova LP, Galan' IK (2003) Material science, process, and service aspects of polyurethane electrically insulating compounds workability. In: Fizichni metody ta zasoby kontrolu setredovishch, materialiv i vyrobiv (Physical methods and means for monitoring environments, materials, and products). No. 8. Karpenko Phys. Mech. Inst., Lviv, p 182–186
17. Marukha VI, Vasilechko VO, Genega BY et al. (2003) Waterproofing cover selection rules for protection of a sewage collector against very aggressive media volleys. In: Proc. int. water forum "Aqua Ukraine 2003", Ukrainian Water Association, Kyiv, Nov. 4–6, 2003, p 194–195
18. Janiak Z, Kapelko A, Noga I, Podolski B (1998) Uwarunkowania materialowe remontu betonowych ŝcian sztolni w elektrowni wodnej (Properties of materials for repairing concrete walls in hydroelectric tunnel). In: Prace Naukowe Instytutu Budownictwa Politechniki Wroclawskej (Sci. Proc. Inst. Constr. Tech. Univ. Wroclaw). No. 28. Tech. Univ., Wroclaw, p 67–74
19. Verbetskiy VG (1976) Prochnost' i dolgovechnost' betona v vodnoy srede (Concrete strength and durability in water environment). Stroyizdat, Moscow
20. Serednitskiy YA, Banakhevich AV, Dragi lєv AV (2008) Protivokorrozionnaya izolyatsiya magistral'nykh gazonefteprovodov (Anticorrosion insulation of gas and oil mains). Splain: Kyiv, Lviv
21. Serednitskiy Y, Banakhevich Y, Dragi lєv A (2004) Suchasna protikoroziyna izolyaciya v truboprovidnomu transporti (Modern anticorrosion insulation in pipeline transport). 2nd Part. Splain: L'viv
22. Kadurina TI, Omel'chenko SI (1980) Hydrolytic stability and protective properties of ester polyurethanes. Paint Lacquer Mater 3:4–6
23. Serednitskiy YA, Kucha ST, Zhuravleva RT et al (1973) Experimentally tested properties of waterproof polyurethane coverings. Phys Chem Mech Mater 5:33–36
24. Lipatov YS, Kercha YY, Sergeeva AM (1970) Struktura i svoistva poliuretanov (Structure and properties of polyurethanes). Nauk. Dumka: Kyiv
25. O'Chaugnessy A, Hoeschale GK (1971) Hydrolytic stability of a new urethane elastomers. Rubber Chem Technol 44:52–61
26. WEBAC® CGmbH. http://www.webac.de. Accessed 22 March 2013
27. Berlin AA, Shutov FA (1978) Penopolimery na osnove reaktsionno-sposobnykh oligomerov (Polymer foams based on reactive oligomers). Stroyizdat: Moscow
28. Bulatov AG (1978) Penopoliuretany v promyshlennosti i stroitel'stve (Polyurethane foams in industry and construction). Mashinostroyeniye: Moscow
29. Bulatov AG (1983) Poliuretany v sovremennoi tekhnike (Polyurethanes in modern technology). Mashinostroyeniye: Moscow
30. Dement'ev AG, Tarakanov OG (1983) Struktura i svoystva penoplastov (Structure and properties of foam plastics). Stroyizdat: Moscow
31. MC-Bauchemie Müller GmbH & Co. KG. (2013) http://www.mc-bauchemie.ru/articles. Accessed 22 March 2013

Chapter 6
Serviceability Estimations for Elements of Building Structures

Abstract Chapter 6 describes analytical models of and solutions to specific problems concerning strength of deformed bodies with defects filled with injection materials. Obtained solutions are the theoretical basis for service life estimations for structural elements after renewal by injection technologies.

For this purpose, a mathematical model of a cracked material healed with injection technologies was developed. The authors analyzed the model in both 2D and 3D formulations. This chapter presents applicability limits of results obtained in two-dimensional approximation. Investigation of crack wedging effects by injection mixtures shows that such effects are significant since they can lead to the growing of the initial crack-like defects under certain conditions.

The extent of filling of defects with an injection material is the important parameter of the technology under consideration. Complete defect filling, in practice, is often hardly attainable, due to various reasons. Therefore, it is important to develop approaches for evaluation of the influence of incomplete defect filling on the effectiveness of damaged structural element renewal.

The problem of injection into a damaged body containing a system of mutually interacting cracks is also considered. The authors consider in detail injection into a system of two cracks. They study the effectiveness of strength restoration for the case of cylindrical structural elements. The solution for transverse compression of a cylindrical element along the planar defect is included. Such specimen configuration is widely used for strength at the testing of brittle materials for strength and fracture toughness. The convenience of such a configuration is in that it requires no special equipment for experiments except for a compression machine. The basic experimental investigations necessary for optimization of injection technologies have been performed using this scheme alone.

Technologies for injecting defect zones in long-term structures are widely used in practice for the renewal of load-carrying capability [1]. However, necessary mathematical models and methods for estimating the serviceability and residual service life (resource) of the renewed structural elements are incomplete at present. The studies in this field are required for the optimization of injection technologies and the prediction of future service life of the restored structures.

One way to develop the required approaches lies in frames of fracture mechanics, the science dealing with the strength of defect bodies, including the bodies containing cracks filled with a foreign material.

V. V. Panasyuk et al., *Injection Technologies for the Repair of Damaged
Concrete Structures,* DOI 10.1007/978-94-007-7908-2_6,
© Springer Science+Business Media Dordrecht 2014

The mathematical theory of deformable bodies with cracks exists to date within the continual model of solid deformable bodies [2–5]. This theory has enabled the derivation of solutions for a wide class of boundary problems related to bodies with cracks. The derived solutions combined with fracture criteria permit quantitative strength estimations for bodies impaired by cracks.

However, in order to estimate the serviceability of solid deformable bodies (concrete elements) with defects such as cracks, voids, loosened areas, etc., a boundary problem of limit equilibrium for the bodies with filled cracks must be solved.

Similar problems compose a separate class of boundary problems in the mechanics of elastic continuum. As opposed to cracks with surfaces, either free of stresses or loaded by known stresses, the filled defects need additional conditions determining the interaction (contact) between host and filler materials. This interaction, reflecting the partial transfer of the load born by the body with a filled crack, represents the essence of the damage-healing concept and determines selection of optimal filling materials subject to injection. The efficiency of such a strengthening effect in a defect-impaired body obviously depends on the adhesion strength between materials adjoining the contact interface. In turn, the nature of the adhesion depends on features of physical and chemical interaction between the materials during solidification or hardening. We shall simulate these bonds with the force factors independent of the nature of the interaction in terms of inhomogeneous continuum mechanics, with the aim of developing the respective mathematical models for estimating the strength of such composite materials and, on this basis, optimize the selection of materials most suitable for injection healing of crack-like defects in structural elements.

6.1 Models of Deformable Bodies with Thin Inclusions

The mathematical models for determining the strength (limit equilibrium state) of solid deformable bodies with thin inclusions were the subject of many studies. Monographs [6–10] present the literature reviews in this field.

The general approach to solving this problem is in the following. A model substitutes the inclusion with a cavity and considers equilibriums of the inclusion and the body with a cavity in place of inclusion as separate problems (Fig. 6.1).

Surface V of the cavity is loaded by, generally speaking, unknown stresses $\sigma_{3j}^{\pm}$ ($j = 1, 2, 3$). Signs $(+)$ and $(-)$ are related, respectively, to the upper $V^{+}(z > 0)$ or lower V^{-} ($z < 0$) parts of the surface V. Numerical indices 1, 2, 3 correspond to the coordinate axes x, y, z. The stresses possess known values only in certain special cases:

a. The inclusion with zero rigidity $\sigma_{3j}^{\pm} = 0$;
b. The case of non-ideal mechanical contacts with known certain components of stress vectors: In particular, in the case of smooth contact $\sigma_{31}^{\pm} = \sigma_{32}^{\pm} = 0$, in the case of inclusion exfoliation in the part of surface V ($V = V^{+} + V^{-}$) all components $\sigma_{3j}^{\pm}$ have known values, etc.

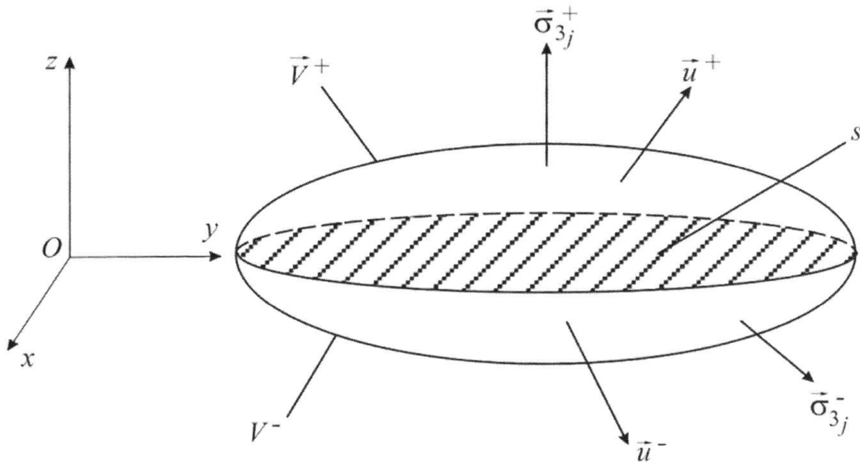

Fig. 6.1 Sketch illustration of inclusion

Assuming the inclusion to be thin, we can suppose the surface V to consist of two plane areas contacting along the plane $z=0$. In this approximation, values $\sigma_{3j}^{\pm}$ are applicable to the middle surface S in the plane $z=0$. As a result, the boundary problem of elasticity theory for a body with a slit in area S loaded by stresses $\sigma_{3j}^{\pm}$ appears.

As known, the slit in a body is the surface of displacement discontinuity. In other words, displacements u_j experience jump $[u_j] = u_j^{+} - u_j^{-}$ at transition over the area S. This boundary problem of elasticity theory has the formal solution if the jumps of displacements $[u_j]$ and stresses $[\sigma_{3j}]$ have known values. In the given case, the solutions of equilibrium equations in the form of Papkovich-Neyber (for three-dimensional problems) or complex potentials of Kolosov-Muskhelishvili (for two-dimensional problems) combined with the Fourier integral conversion method are effective. These solutions yield dependences enabling the calculation of stresses and displacements in the deformed body starting from the jumps in area S:

$$u_j = u_j([u_j], [\sigma_{3j}]); \quad \sigma_{ij} = \sigma_{ij}([u_j], [\sigma_{3j}]). \tag{6.1}$$

Consequently, if the jumps $[u_j]$ and $[\sigma_{3j}]$ are known, then the problem of determining stresses and strains in the inhomogeneous body is solved.

A general case of inhomogeneity in a body includes six unknown jumps of displacements and stresses that require additional relationships for determination. They include inclusion equilibrium equations plus relationships between stresses and displacements in the inclusion.

Having applied an averaging operation across the inclusion's thickness to these relationships, we obtain six constraints for displacements and stresses on the interface V:

$$f_n(u_j^{\pm}, \sigma_{3j}^{\pm}) = 0, n = 1, 2 \ldots 6. \tag{6.2}$$

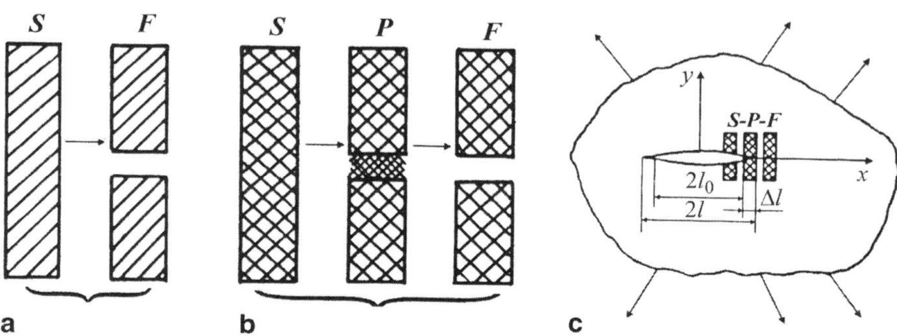

Fig. 6.2 Schemes of material elements fracture: classic (**a**), non-classical (**b**), non-classical near crack tip (**c**)

Substitution of expressions for displacements and stresses via jumps $[u_j^\pm]$, $[\sigma_{3j}^\pm]$ in accordance with relationships (6.1) instead of $u_j^\pm$, $\sigma_{3j}^\pm$ in (6.2) results in the system of six integro-differential equations with unknown jumps $[u_j^\pm]$, $[\sigma_{3j}^\pm]$ ($j = 1, 2, 3$):

$$g_n([u_j^\pm], [\sigma_{3j}^\pm]) = 0, n = 1, 2 \ldots 6.$$

The number of equations can be less dependent on the nature of the inclusion, the conditions of contact with a host material, and the applied loading mode. In some cases, the system even reduces to a single equation.

6.2 Material Fracture Criteria in Fracture Mechanics

Developed at the beginning of the 20th century, fracture criteria and methods for estimating the strength of materials and structures is based on the continuum as an approximated model of a solid body with given rheological properties (in particular, the elastic continuum). An element of the deformed body can occur in one of two states (Fig. 6.2): solid (S-state) or fractured (F-state). The material converges from S-state into F-state, i.e., fails instantly once the stress state, established in frames of the adopted rheological model, reaches a certain critical magnitude (for example, once the tensile stress in a given point of a deformed solid body reaches the tensile strength value σ_B).

However, the above classic approach fails to estimate the strength of materials containing acute crack-like defects. The cause is in the tendency of the classic approach of continuum mechanics to ignore the peculiar stress state near a crack-like acute defect tip in the loaded body when estimating the strength (fracture stress) of the material.

The fundamental idea of the non-classical approach (fracture mechanics of materials) is the following [4] (Fig. 6.2). Assume that an element of a deformed body converges from S-state (solid) to F-state through some intermediate P-state that one must take into consideration while solving the strength problem for the body with crack-like defects.

The P-states (pre-failure zones) arise in those areas of the deformed body where deformation of the material is beyond the limit of elasticity. Such areas are remarkable for the most intensive plastic flow, interaction with surrounding liquid or gaseous environment, diffusional processes, accumulation of microdamages, etc., which cause local fracture, i.e., transitions $S{\to}P{\to}F$ (Fig. 6.2).

Energy Fracture Criterion

Griffith [11] was the first to relate the strength of a solid deformable body (or material) with the crack present in it mathematically. For this, he applied an energy criterion, assuming that the crack can propagate, i.e., the equilibrium state of the deformed body reaches its limit, when the elastic energy release rate W during crack growth becomes equal to the surface energy U required for creation of new surfaces in this body.

Having applied this principle to the problem of a plate with a crack under tension, Griffith derived the following equation:

$$\sigma_c = \sqrt{\frac{2E\gamma}{\pi l}}, \tag{6.3}$$

where σ_c is the critical (fracture) load value (Fig. 6.3); E is the Young modulus of the material; γ is a specific fracture energy identified by Griffith with the specific surface energy of the material, and $2l$ is the crack length.

Equation (6.3) has a key importance in both theory and practice. It shows that the strength of a crack containing material depends on the physical and mechanical constants of the material and crack length. At the same time, it indicates that the strength of a body can become higher by either diminishing the defect size (crack length l) or hindering its propagation (growth). One way to effect both of the above factors is in healing, that is, filling the crack or defect with a foreign material.

Irwin Criterion

The next important milestone in the development of fracture mechanics emerged in connection with studying stress fields near the crack tip in the deformable solid body [12]. Such studies established that the strain distribution around the crack tip is representable as a superposition of strain fields belonging to three modes (Fig. 6.4) corresponding to three kinds of crack edge displacement: normal cleavage (I), transverse shear (II), and longitudinal shear (III), whereas components of stress tensor σ_{ij} (Fig. 6.5a) near the crack tip have the form:

$$\sigma_{ij} = \frac{1}{\sqrt{2\pi r}}\left\{K_{\mathrm{I}}f_{1ij}(\theta) + K_{\mathrm{II}}f_{2ij}(\theta) + K_{\mathrm{III}}f_{3ij}(\theta)\right\} + O(1), \tag{6.4}$$

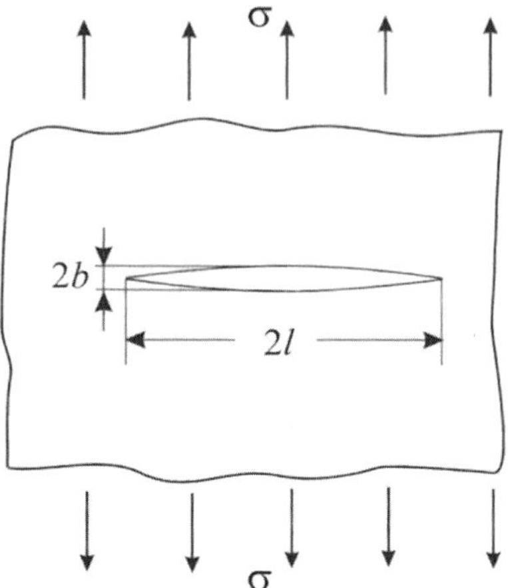

Fig. 6.3 The plate with a crack under tension

where $K_I = K_I(p, l)$, $K_{II} = K_{II}(p, l)$, $K_{III} = K_{III}(p, l)$ are the stress intensity factors (SIF) being functions of body geometry, crack dimensions, and applied load (p); $f_{kij}(\theta)$ are known functions ($k = 1,2,3$); $O(1)$ is a limited quantity at $r \to 0$; $i, j = 1, 2, 3$. Indices 1, 2, 3 correspond to coordinate axes x, y, z.

Taking into consideration expressions (6.4), G. Irwin [12] formulated in 1957 the criterion of the limit equilibrium state of a deformable body with a crack under quasistatic loading for the case when the fracture occurs in pure cleavage, i.e., $K_I \neq 0$ while $K_{II} = 0$ and $K_{III} = 0$. This criterion implies that the following statement is valid: The applied load p is limit equilibrium ($p = p_*$), if SIF $K_I (p, l)$ for a deformed body with a crack is equal to a certain material constant K_{IC}, that is

$$K_I(p_*, l) = K_{IC}. \tag{6.5}$$

G. Irwin proved that in the case of brittle fracture when $\Delta l \ll l_0$ (see Fig. 6.2c) criterion (6.3) is equivalent to Griffith's energy criterion.

In the 1960s, the authors [13] generalized the criterion (6.5) for the case when a plate with a crack is in a complex (two-dimensional) stress state, that is, when

$$K_I(p, l) \neq 0, \; K_{II}(p, l) \neq 0, \; K_{III}(p, l) = 0$$

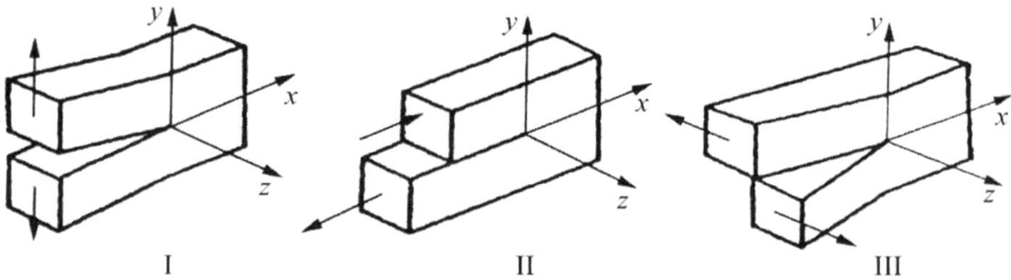

Fig. 6.4 Three modes of crack edge primary displacement near the crack tip

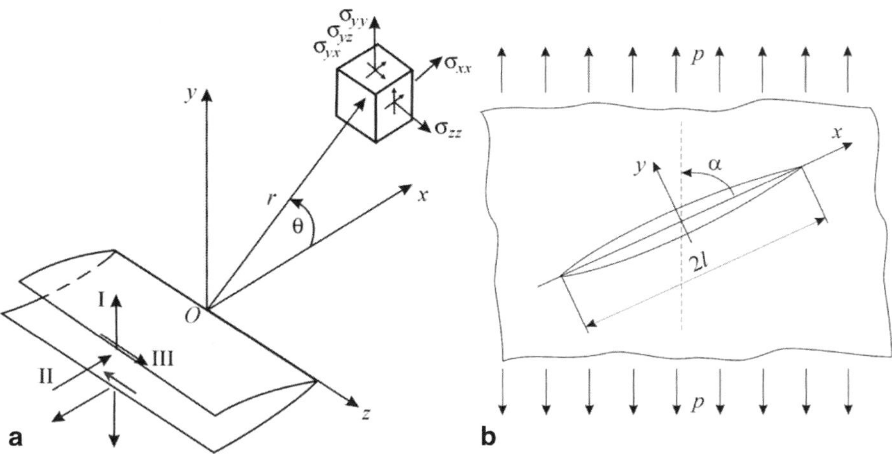

Fig. 6.5 Local coordinate system near the crack front (*line Oz*) and components (*I, II, III*) of the crack edges displacement vector (**a**), tension of a plate with an arbitrarily oriented crack (**b**)

The authors adopted the following hypothesis as the condition determining the direction of initial crack propagation (σ_θ-criterion [13]): The crack propagates in those planes $\theta = \theta_*$ (the plane zOr in Fig. 6.5a), in which the normal tensile stress (σ_θ) intensity factor has the highest values, that is

$$\sigma_\theta(r,\theta) = \frac{F(K_\mathrm{I}, K_\mathrm{II})}{\sqrt{r}} + o(1), \ \lim_{r \to 0} \sqrt{r} \left. \frac{\partial \sigma_\theta(r,\theta)}{\partial \theta}\right|_{\theta=\theta_*} = 0.$$

The above approach first enabled the derivation of the limit equilibrium equation for the deformed body (plate) with an arbitrarily oriented crack [13] (Fig. 6.5b).

$$\cos^3 \frac{\theta_*}{2}[K_\mathrm{I}\left(p_*, \alpha, l\right) - 3\tan \frac{\theta_*}{2} K_\mathrm{II}(p_*, \alpha, l)\,] = K_{\mathrm{I}C}, \qquad (6.6)$$

$$\theta_* = 2\mathrm{arctg}^3 \frac{K_\mathrm{I} - \sqrt{K_{\mathrm{I}0}^2 - K_{\mathrm{II}0}^2}}{4 K_{\mathrm{II}0}^2}, \ K_\mathrm{I} > 0,\ K_\mathrm{II} \neq 0, \qquad (6.7)$$

$$K_\mathrm{I} = p\sqrt{\pi l}\sin^2\alpha, \ K_\mathrm{II} = p\sqrt{\pi l}\sin\alpha\cos\alpha, \qquad (6.8)$$

where θ_* is the angle of initial crack growth; α is the angle of crack orientation relative to the direction of loading p.

Equations (6.6) and (6.7) were confirmed experimentally [14] and are used to establish the limit equilibrium (strength) of cracked bodies under a complex stress state.

The mathematical models of Griffith and Irwin are the basis of so-called linear mechanics of brittle fracture in solid bodies. The main distinctive feature of this branch of fracture mechanics consists of the size of the pre-failure zone near the

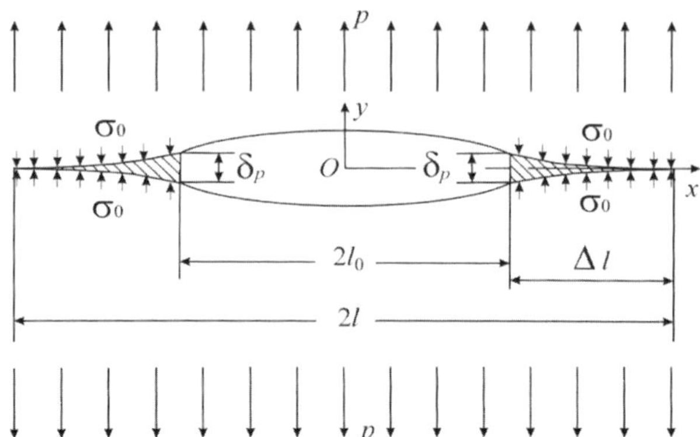

Fig. 6.6 Geometry of δ_c-model

crack tip with material deformed beyond the limit of elasticity, small when compared with the crack and/or body sizes. The stress state in such a body follows from solving the respective problems of linear elastic theory for the body with a crack or slit.

These solutions result in calculation of SIF (K_{i0} (p, l), $i = $ I, II, III). Crack growth resistance of the material, or, in other words, fracture toughness, is estimated by experimentally measured values of K_{IC}, K_{IIC}, and K_{IIIC} [15].

Deformational Criterion

If the above conditions of linear fracture mechanics fail, i.e., the plastic zone (the area of P-states) is comparable in size with the crack length, another criterion is required to determine the limit equilibrium state for the body with a crack. Such a role belongs to the deformational criterion based on the deformational characteristics of the material, in particular, the critical crack opening (CCO-criterion) [16]. According to this criterion, the limit equilibrium state for the elasto-plastic body with a crack appears when the crack opening in the crack tip δ_p (Fig. 6.6) reaches the critical value δ_{IC} subject to experimental measuring. Therefore, the deformational criterion has the form

$$\delta_p(p_*) = \delta_{\mathrm{IC}}, \tag{6.9}$$

where $p = p_*$ is the critical (highest allowable) load value.

This criterion, as distinct from the previous, has no limitations for the plastic zone (zone of P-states) size, and, hence, is a basis of non-linear (elasto-plastic) fracture mechanics. The authors [16–18] first presented a mathematical model (δ_c-model) of a body with a crack enabling the calculation of values $\delta_{\mathrm{I}}(p)$. The references [19], [20] contain more information on this and similar approaches. The δ_c-model presents the pre-failure area (zone of P-states) Δl as an additional slit having interacted (attracted) with force σ_0 at opposite edges (Fig. 6.6). The force σ_0 for the ideal elasto-plastic material is supposed equal to $\sigma_0 = \sigma_T$ (where σ_T is a plastic yield point of the material). If the material has a non-zero strain hardening, the value σ_T is determined from the stress-strain curve for a plain specimen and thereafter value σ_0 is either accepted as average between the yield point and strength point $\sigma_0 = (1/2)(\sigma_T + \sigma_B)$ or calculated using the formula:

$$\sigma_0 = \sigma_T + E(\varepsilon_c - \varepsilon_T) - \sqrt{E^2(\varepsilon_c - \varepsilon_T)^2 - 2AE}, \quad A = \int\limits_{\varepsilon_c}^{\varepsilon_T} [f(\varepsilon) - \sigma_T]d\varepsilon,$$

where $f(\varepsilon)$ is the stress-strain curve in the plastic strain stage; ε_T is the strain at the yield point, and ε_c is the ultimate strain of the material.

The deformational criteria for modes of transverse (δ_{II}) or longitudinal (δ_{III}) shear have a form similar to the δ_{IC}-criterion with the shearing interaction stresses τ_{20} or τ_{30} between the opposite slit or model crack sides in the pre-failure zone instead of normal ones [4].

Let us consider, for instance, the solution of the problem of the limit equilibrium state in the deformed by a tension plate having a central crack with length $2l_0$ in terms of the δ_c-model [16] (Fig. 6.6).

Let us adopt the Cartesian coordinate system xOy ($z=0$) with the origin in the crack center and the axis x directed along the crack so that the crack occupies the axis segment $(-l_0, l_0)$. Tensile forces are applied in infinity (far from the crack) so that $\sigma_x = 0$, $\sigma_y \infty = p$. Let us replace the plastic zones near the crack tips (P-states of material) with additional slots with length $\Delta l = d_p$ loaded by attractive forces σ_0 acting between opposite edges.

In such a statement, the above problem is the elastic theory problem for an infinite plate (xOy) with a crack/slit and the following boundary conditions:

- in infinitely remote points of the plate, stresses (forces) are given as follows:

$$\sigma_{xx}^{(\infty)} = 0, \quad \sigma_{yy}^{(\infty)} = p, \quad \sigma_{xy}^{(\infty)} = 0, \tag{6.10}$$

- at the crack contour $(-L \leq x \leq L)$, where $L = l_0 + d_p$:

$$\sigma_{xy}(x,0) = 0, \quad \sigma_y(x,0) = \begin{cases} 0 & \text{for } |x| < l_0, \\ \sigma_0 & \text{for } l_0 \leq |x| \leq L. \end{cases} \tag{6.11}$$

Solution of the elastic theory problem for the infinite plate with boundary conditions (6.10), (6.11) gives the formula for calculating the opening between crack edges [16], [17]. This formula in crack tips, that is, in points with coordinates $(\pm l_0, 0)$, has the following form:

$$2u_y(p, \pm l_0, 0) = -8l_0\sigma_0 c \ln \cos \frac{\pi p}{2\sigma_0}, \tag{6.12}$$

where $c = 1/(\pi E)$ for a generalized plane stress state (thin plate); $c = (1 - v^2)/(\pi E)$ for plane strain (thick plate); E is the Young modulus, and v is the Poisson ratio for material of the plate.

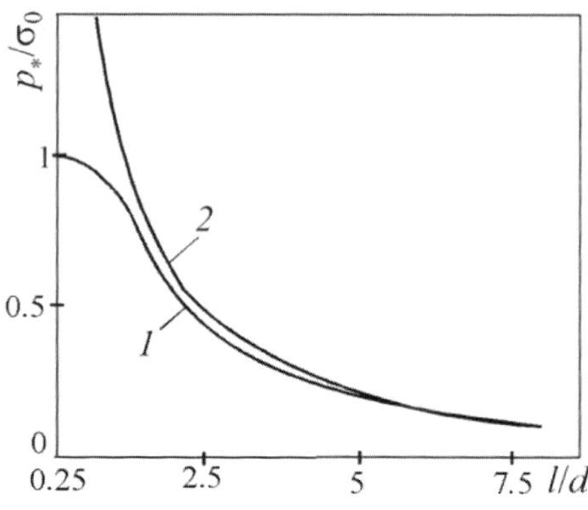

Fig. 6.7 Limiting load dependence on the crack length according to formula (6.13) (curve *1*) or Griffith equation (curve *2*)

Using formula (6.12) and criterion (6.9), we obtain the following expression for estimating the critical load $p = p_*$:

$$\delta_{\mathrm{IC}} = 2u_y(p_*, \pm l_0, 0) = -8l_0\sigma_0 c \ln \cos \frac{\pi p_*}{2\sigma_0},$$

from whence

$$p_* = \frac{2\sigma_0}{\pi} \arccos e^{-\eta}, \quad \eta = \frac{\delta_{\mathrm{IC}}}{8l_0\sigma_0 c}. \tag{6.13}$$

Expanding the function $\arccos e^{-\eta}$ into a series with small parameter $\eta \ll 1$, neglecting all η powers except the first, and taking into account the relation $2\gamma = \delta_{\mathrm{IC}}\sigma_0$, which is basic for the δ_c-model, we come to the Griffith equation (6.3):

$$p_* = \sqrt{\frac{2E\gamma}{\pi l}}.$$

However, it follows from the Griffith equation that p_* infinitely grows at $l_0 \to 0$, which disagrees with the experimental data. On the other hand, the formula (6.13) predicts that p_* approaches σ_0 at $l_0 \to 0$, i.e., the load value stays finite. This physically obvious result could hardly follow from other generalizations of Griffith's concept (see Fig. 6.7).

6.3   Strength of a Body with a Filled Crack

Problem Definition

Let us consider a massive structural element (deformable solid body) with an inner crack-like defect.

Assume that the defect is far from the body surface. In such an assumption, the mathematical model of the body can represent an elastic continuum with an inner

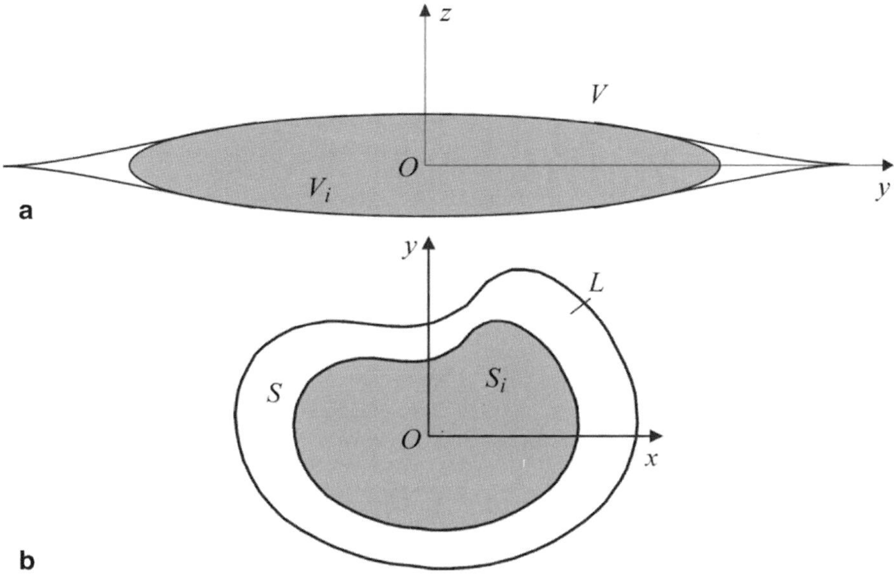

Fig. 6.8 Sectional view of a body with the crack in planes $z=0$ (**a**) and $x=0$ (**b**)

crack of a given configuration. Let us establish the Cartesian coordinate system $Oxyz$ (Fig. 6.8) such that the crack with surface V occupies in the plane $z=0$ a certain area S confined inside a smooth contour L. The crack configuration in plane $x=0$ is shown in Fig. 6.8a.

In addition, the present problem definition should take into account the following. When the crack undergoes filling by a foreign material by means of injection technology, this material is initially in a liquid state and then solidifies and forms bonds with the host body.

Therefore, situations are possible in which the injected material after solidification (polymerization or crystallization) occupies only part V_i of the crack volume and part S_1 of the crack surface (Fig. 6.8) while sticking together the crack edges and strengthening the damaged body.

Thus, the problem of stress state in a body with a filled crack transforms to the elastic theory problem for a piecewise uniform deformable body. Host-guest material interface can have different properties, including both ideal and non-ideal mechanical contacts with exfoliation, slip, etc.

These properties must be defined in order to solve the respective elastic theory problem for a piecewise uniform deformable body and find stress field components and strains near such a defect (the filled crack).

In the literature, many solutions of similar problems are described [6–10]. Basic (key) characteristics of the sought field for the acute crack-like defects are so-called SIF, which depend on the geometry of a body under study, crack shape, and hardness of the filler, as well as the crack edge displacement mode (see Fig. 6.4) near the crack tip or crack front, namely:

$$K_{\mathrm{I}} = K_{\mathrm{I}}(p,l),\, K_{\mathrm{II}} = K_{\mathrm{II}}(p,l),\, K_{\mathrm{III}} = K_{\mathrm{III}}(p,l), \tag{6.14}$$

where p is loading intensity and l is the typical linear crack size connected with its configuration V.

In order to find SIF (K_I, K_{II}, K_{III}), it is necessary to define and solve the respective boundary problem of empty or filled cracks in a deformable solid body.

The SIF values found for a specific problem determine the highest admissible load $p = p^*$. Loading the body below this value preserves its integrity, i.e., strength, although a small load increment beyond this value causes crack propagation and fracture.

One can calculate the limit equilibrium values of an applied load from the respective criteria:

$$K_I(p_*, l) = K_{IC}, \text{if} K_I \neq 0, \ K_{II} = 0, \ K_{III} = 0; \tag{6.15}$$

$$K_{II}(p_*, l) = K_{IIC}, \text{if} K_I = 0, \ K_{II} \neq 0, \ K_{III} = 0; \tag{6.16}$$

$$K_{III}(p_*, l) = K_{IIIC}, \text{if} K_I = 0, \ K_{II} = 0, \ K_{III} \neq 0. \tag{6.17}$$

In the case of a complex stress state near the crack front, i.e., when $K_I \neq 0$, $K_{II} \neq 0$, $K_{III} \neq 0$, there is a complicated functional for determining the limit equilibrium state of a body with a crack-like defect in the form:

$$F(K_I, K_{II}, K_{III}, K_{IC}, K_{IIC}, K_{IIIC}) = 0. \tag{6.18}$$

Values K_{IC}, K_{IIC}, K_{IIIC} are subject to experimental measurement.

Finally, the obtained results enable estimation of the efficiency of serviceability restoration for the structural element impaired by crack-like defects (acute voids) by comparing values of ultimate load p_* for empty or filled cracks. Such a comparison allows for choosing the best materials for implementation of injection technologies.

Boundary Conditions for Bodies with Filled Cracks

Let us ideally remove the filled material from the crack and replace it with stresses on the crack or inclusion contour (boundary). Assume that these stresses have the form [6]:

$$\vec{\sigma}_z^* \left\{ \sigma_{xz}^*, \sigma_{yz}^*, \sigma_{zz}^* \right\} = \vec{\sigma}_z^* \left\{ \frac{[u_x^*]}{2h} \mu_1, \ \frac{[u_y^*]}{2h} \mu_1, \ \frac{[u_z^*]}{2h} E_1 \right\}. \tag{6.19}$$

Here μ_1 and E_1 are the shear and Young moduli of the filled material; u_x^*, u_y^*, u_z^* are the components of the displacement vector for inclusion surface points V; square brackets denote a function jump at the contour S; $2h$ is the inclusion thickness or crack opening.

Relations (6.19) between stresses and displacements on the crack surface adequately describe interaction of the filler with its host material, if the filler's hardness is less than the hardness of the matrix, that is, valid, e.g., in the case of concrete as the host material and a polymer as the filler.

Using the superposition principle valid in the linear elastic theory, we can express the stress state in the body as the sum of two components, the stress state of a homogeneous (containing no cracks) body under applied load and the stress state of the body with a crack loaded by application to its surfaces forces $\vec{\sigma}_z = \vec{\sigma}_z^* - \vec{e}_3 \widehat{\sigma}^0$ on S_i and $\vec{\sigma}_z = -\vec{e}_3 \widehat{\sigma}^0$ on $S - S_i$. (Here $\vec{e}_3$ is a unit direction vector along axis z).

Then, the displacement vector $\vec{u}_z^*$ for the surface points becomes a sum of two components, too:

$$\vec{u}^* = \vec{u}^0 + \vec{u}, \tag{6.20}$$

where $\vec{u}_z^0$ are displacements of surface points V in the homogeneous body and $\vec{u}$ are displacements of crack edges under action of forces $\vec{\sigma}_z = \vec{\sigma}_z^* - \vec{e}_3 \widehat{\sigma}^0$ applied in area S_i and $\vec{\sigma}_z = -\vec{e}_3 \widehat{\sigma}^0$ applied in area $S - S_i$.

Taking into account the small thickness of inclusion V, let us translate the boundary conditions to median area S. Thereby we reduce the problem to the elastic theory boundary problem for an infinite body with mathematical slit S loaded by the following surface stresses:

$$\vec{\sigma}_z = \begin{cases} \vec{\sigma}_z^* - \vec{e}_3 \widehat{\sigma}^0, & (x, y) \in S_i, \\[2mm] -\vec{e}_3 \widehat{\sigma}^0, & (x, y) \in (S - S_i). \end{cases} \tag{6.21}$$

In absence of bulk forces, one can reduce the solution of an elastic theory problem to determination of displacement vector $\vec{u}$ from equilibrium equations:

$$\mathrm{grad\ div\ } \vec{u} + (1 - 2v)\Delta \vec{u} = 0 \tag{6.22}$$

and respective boundary conditions (in given case conditions (6.21)), where $\Delta = \partial/\partial x^2 + \partial/\partial y^2 + \partial/\partial z^2$ is the Laplace operator and v is the Poisson ratio.

Let us apply the general solution of Eq. (6.22) expressed in three Papkovich-Neyber harmonic functions Φ_1, Φ_2, and Φ_3 [21]:

$$2\mu \vec{u} = z \mathrm{grad\ div\ } \vec{\Phi} - 2(1 - 2v)\frac{\partial \vec{\Phi}}{\partial z} + (1 - 2v)(\mathrm{grad\ } \Phi_3 - \vec{e}_3 \mathrm{div\ } \vec{\Phi}), \quad \vec{\Phi} \tag{6.23}$$

$$= \vec{\Phi}\,\{\Phi_1, \Phi_2, \Phi_3\}$$

Stress tensor $\hat{\sigma}$ is expressible in terms of the displacement vector based on Hook's law and Cauchy relations:

$$\hat{\sigma} = \mu \left(\frac{2v}{1 - 2v}(\vec{\nabla}\vec{u}) \right) \hat{I} + \vec{\nabla}\vec{u} + (\vec{\nabla}\vec{u})^*, \tag{6.24}$$

Where $\hat{I}$ is the unit tensor; $\vec{\nabla}$ is the Hamilton operator and $(\vec{\nabla}\vec{u})^*$ is the tensor transposed from tensor $(\vec{\nabla}\vec{u})$ [21].

From Eqs. (6.23), (6.24), the stress vector in a plane $z = \mathrm{const}$ has the form:

$$\vec{e}_3\hat{\sigma} = z \, \mathrm{grad} \, \mathrm{div} \frac{\partial \vec{\Phi}}{\partial z} - \frac{\partial^2 \vec{\Phi}}{\partial z^2} + v \left(\vec{e}_1 \frac{\partial}{\partial y} - \vec{e}_2 \frac{\partial}{\partial x} \right) \left(\frac{\partial \Phi_2}{\partial x} - \frac{\partial \Phi_1}{\partial y} \right). \qquad (6.25)$$

Let us represent the harmonic function vector $\vec{\Phi}$ in the form of an integral Fourier series:

$$\vec{\Phi}(x, y, z) = \int\limits_{-\infty}^{\infty}\!\!\int \vec{A}(\xi, \eta) \exp\left(-z\sqrt{\xi^2 + \eta^2} + i(x\xi + y\eta)\right)\frac{d\xi \, d\eta}{\sqrt{\xi^2 + \eta^2}}, \qquad (6.26)$$

where $\vec{A}(\xi, \eta)$ is the unknown vector function subject to determination.

The displacement vector continuity beyond area S yields us with an account of Eq. (6.23):

$$\int\limits_{-\infty}^{\infty}\!\!\int \vec{A}(\xi, \eta) \exp\left(i(x\xi + y\eta)\right)d\xi \, d\eta = 0, (x, y) \in \bar{S}. \qquad (6.27)$$

Let us designate:

$$\frac{1 - v}{\mu} \int\limits_{-\infty}^{\infty}\!\!\int \vec{A}(\xi, \eta) \exp\left(i(x\xi + y\eta)\right)d\xi \, d\eta = \begin{cases} \vec{u}, & (x, y) \in S \\ 0, & (x, y) \in \bar{S} \end{cases}. \qquad (6.28)$$

The inverse Fourier transform gives us the sought vector function $\vec{A}(\xi, \eta)$ from Eq. (6.28) in the form:

$$\vec{A}(\xi, \eta) = \frac{\mu}{4\pi^2(1 - v)} \int\limits_{S}\!\!\int \vec{u}(x, y) \exp\left(-i(x\xi + y\eta)\right)dx \, dy. \qquad (6.29)$$

After some conversions, substitution of this relationship for $\vec{A}(\xi, \eta)$ into Eq. (6.23) results in vector function $\vec{\Phi}$ expressed through area S points displacements:

$$\vec{\Phi}(x, y, z) = \frac{\mu}{2\pi(1 - v)} \int\limits_{S}\!\!\int \frac{\vec{u}(x_1, y_1)dx_1 dy_1}{\sqrt{z^2 + (x - x_1)^2 + (y - y_1)^2}}. \qquad (6.30)$$

The boundary conditions (6.21) combined with Eqs. (6.19), (6.28), (6.30) form a system of integral equations for determining displacements $\vec{u}(x, y)$:

$$
\begin{cases}
\nabla^2_{xy} \iint_S \dfrac{u_x(\xi, \eta)\, d\xi\, d\eta}{\sqrt{(x - \xi)^2 + (y - \eta)^2}} \\[2mm]
\quad + v \left(\dfrac{\partial^2}{\partial x\, \partial y} \iint_S \dfrac{u_y(\xi, \eta)\, d\xi\, d\eta}{\sqrt{(x - \xi)^2 + (y + \eta)^2}} - \dfrac{\partial^2}{\partial y^2} \iint_S \dfrac{u_x(\xi, \eta)\, d\xi\, d\eta}{\sqrt{(x - \xi)^2 + (y - \eta)^2}} \right) \\[2mm]
\quad = \dfrac{4\pi(1 - v^2)}{E}
\begin{cases}
\dfrac{[u_x^0 + u_x]}{2h}\mu_1 - \sigma_{xz}^0, & (x, y) \in S_1 \\[2mm]
-\sigma_{xz}^0, & (x, y) \in S - S_1
\end{cases} \\[4mm]
\nabla^2_{xy} \iint_S \dfrac{u_y(\xi, \eta)\, d\xi\, d\eta}{\sqrt{(x - \xi)^2 + (y - \eta)^2}} \\[2mm]
\quad + v \left(\dfrac{\partial^2}{\partial x\, \partial y} \iint_S \dfrac{u_x(\xi, \eta)\, d\xi\, d\eta}{\sqrt{(x - \xi)^2 + (y - \eta)^2}} - \dfrac{\partial^2}{\partial x^2} \iint_S \dfrac{u_y(\xi, \eta)\, d\xi\, d\eta}{\sqrt{(x - \xi)^2 + (y - \eta)^2}} \right) \\[2mm]
\quad = \dfrac{4\pi(1 - v^2)}{E}
\begin{cases}
\dfrac{[u_y^0 + u_y]}{2h}\mu_1 - \sigma_{yz}^0, & (x, y) \in S_1 \\[2mm]
-\sigma_{yz}^0, & (x, y) \in S - S_1
\end{cases} \\[4mm]
\nabla^2_{xy} \iint_S \dfrac{u_z(\xi, \eta)\, d\xi\, d\eta}{\sqrt{(x - \xi)^2 + (y - \eta)^2}} = \dfrac{4\pi(1 - v^2)}{E}
\begin{cases}
\dfrac{[u_z^0 + u_z]}{2h}E_1 - \sigma_{zz}^0, & (x, y) \in S_1 \\[2mm]
-\sigma_{zz}^0, & (x, y) \in S - S_1
\end{cases}
\end{cases}
\tag{6.31}
$$

After solving the equation system (6.31), one can directly find SIF (K_{I}, K_{II}, K_{III}) through components of the displacement vector $\vec{u}$.

For this, one can remember the fact that local displacements near the crack tip in the three-dimensional case have the same functional structure as displacements in the respective combination of plane stress and longitudinal shear [22]. Accounting for the above-mentioned, we obtain the following expressions for SIF:

$$
K_{\mathrm{I}} = \lim_{n \to 0} \sqrt{\frac{\pi}{2n}} \frac{\mu u_z}{1 - v},
$$

$$
K_{\mathrm{II}} = \lim_{n \to 0} \sqrt{\frac{\pi}{2n}} \frac{\mu u_n}{1 - v},
\tag{6.32}
$$

$$
K_{\mathrm{III}} = \lim_{n \to 0} \sqrt{\frac{\pi}{2n}}\, \mu u_t.
$$

Here n, t are axes of a local Cartesian coordinate system with origin in the contour of area S chosen as shown in Fig. 6.9.

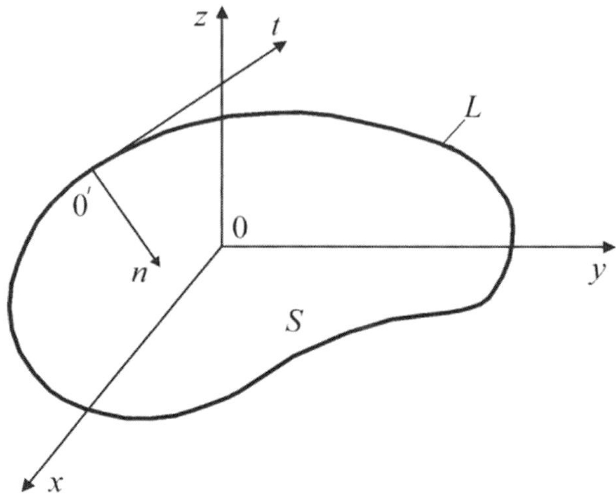

Fig. 6.9 Local coordinate system on the contour of inclusion median area

Strength of Plates with Empty or Filled Cracks

Let us consider the classic Griffith problem for the strength of a plate with a crack (see Fig. 6.3) under tension by forces with intensity p applied far from the crack plane. Both energy [11] and force approaches [12] for such a loading mode yield the plate's strength in the form [11], [12]:

$$p_c = \sqrt{\frac{2\gamma E}{\pi(1 - \nu^2)l}} = \frac{K_{1C}}{\sqrt{\pi l}}. \qquad (6.33)$$

Here γ is the specific fracture energy; E is the Young modulus; ν is the Poisson ratio; K_{1C} is the fracture toughness of the material (critical SIF).

Now, let us determine the limit equilibrium state (strength) of the plate with a crack filled with an injection material. We shall assume that after solidification of the introduced liquid paste in the crack, the bonds form in the interface of materials ensuring the perfect mechanical contact. Besides, we shall suppose that the crack preserves its length under pressure of the injection material.

In order to solve this problem, let us use the model relationships (6.19). Then, the problem of uniaxial tension of a plate with a filled crack will reduce to the problem for the body with a crack loaded by mutually attractive forces on its edges:

$$\sigma_y^*(x) = \frac{[u_y^*(x)]}{2h(x)} E_1, \qquad (6.34)$$

where $[u_y^*(x)]$ is the displacement jump at surfaces of the filled crack; $2h(x)$ is the filler's material thickness and E_1 is the Young modulus of the injection material after hardening.

The definition of interaction between the injection and host materials in form (6.34) reduces the problem to the following boundary problem of crack theory for an infinite plate with a slit $2l$:

$$\sigma_y(x) = -p + \frac{[u_y^*(x)]}{2h(x)} E_1, \quad \sigma_{xy}(x) = 0, \quad |x| \leq l. \tag{6.35}$$

In infinitely remote points of the plate, we have:

$$\sigma_{xx}^\infty = 0, \quad \sigma_{yy}^\infty = 0, \quad \sigma_{xy}^\infty = 0.$$

It follows from Eqs. (6.31) that function $[u_y(x)]$ must result from the singular integro-differential equation [6]:

$$\int \frac{[u_y'(x)]}{t - x} dt = \frac{4\pi(1 - v^2)}{E} \left(p(\omega - 1) + \frac{[u_y(x)]}{2h(x)} E_1 \right), \quad |x| \leq l. \tag{6.36}$$

We had taken into consideration here that $u_y = u_y^* - u_y^0$; $u_y^0 = ph(x)/E$; $\omega = E_1/E$ in accordance with Eq. (6.17).

Equation (6.36) has an exact solution, if the contour of the opened crack/slit has the shape of an ellipse, i.e., $h(x) = b\sqrt{1 - x^2/l^2}$ [6]:

$$[u_y(x)] = \frac{4p(1 - \omega)\sqrt{l^2 - x^2}}{E(1 + 2\beta\omega)}, \quad \beta \equiv \frac{l}{b}. \tag{6.37}$$

Then, stresses in inclusion are

$$\sigma = \frac{p(1 + 2\beta)\omega}{1 + 2\beta\omega}. \tag{6.38}$$

SIF for the filled crack is equal:

$$K_I = \frac{p\sqrt{\pi l}(1 - \omega)}{1 + 2\beta\omega}. \tag{6.39}$$

The expressions (6.38), (6.39) give an opportunity to find the strength of the plate with the filled crack.

Namely, let σ_B^* be the ultimate strength of the injection material.

Since the injection material, according to (6.38), occurs under uniaxial tension along the entire length of the filled crack, the strength theory indicates the following applied loads corresponding to fracture of the filling material:

$$p_c^* = \frac{\sigma_B^*(1 + 2\beta\omega)}{(1 + 2\beta)\omega}. \tag{6.40}$$

Using Eq. (6.39) and the crack propagation criterion $K_1(p_c, l) = \tilde{K}_{IC}$, one can easily calculate the load causing immediate growth of the injected crack:

$$p_c^* = \frac{\tilde{K}_{IC}(1 + 2\beta\omega)}{\sqrt{\pi l}(1 - \omega)}. \tag{6.41}$$

Here $\tilde{K}_{IC}$ is a constant characterizing the resistance of the host material to crack growth near the filled defect.

It is clear that using injection as the method for strengthening a cracked structural element is feasible only in the cases when the injected material will withstand applied loads higher than the limit value $p_c = K_{IC}/\sqrt{\pi l}$.

As Eq. (6.40) shows, the above condition is true if:

$$\sigma_B^* > \frac{K_{IC}\omega(1+2\beta)}{\sqrt{\pi l}(1+2\beta\omega)}. \tag{6.42}$$

Respectively, the optimal injection material must fail later than the host material, that is

$$p_c^* > p_c.$$

From here, taking into consideration Eqs. (6.40) and (6.41), we obtain the estimation or required injection material strength in the form:

$$\sigma_B^* > \frac{\tilde{K}_{IC}\omega(1+2\beta)}{\sqrt{\pi l}(1-\omega)}. \tag{6.43}$$

Under this condition, one can estimate the strength of the plate with the filled crack from Eq. (6.41).

Thus, using the relationships

$$\frac{p_c}{\sigma_B^*} = \chi, \quad \chi \equiv \frac{\tilde{K}_{IC}(1+2\beta\omega)}{\sigma_B\sqrt{\pi l}(1-\omega)}, \quad \beta \equiv \frac{l}{b}, \quad \omega \equiv \frac{E_1}{E}, \tag{6.44}$$

one can estimate the goodness of load-carrying capability renewal for the plate with the filled crack.

Here, the case $\chi = 1$ corresponds to complete strength restoration whereas the case $\chi = \chi_0 = K_{IC}/\sigma_B\sqrt{\pi l}$ corresponds to the non-injected crack. In other words, χ varies within the range from 1 to χ_0.

Equation (6.41) means that one can reach the necessary strengthening by either elevating the parameter β, i.e., injecting at the point of the lowest crack opening, or enhancing the parameter ω that characterizes filler hardness.

To confirm the dependence (6.44), we experimentally tested specimens of concrete as a host material and polyurethane as an injection material.

Cylindrical specimens $100 \times 100 \times 100$ mm with a central through slit imitating a crack were subjected to transverse compression along the crack plane in a uniaxial machine.

This test configuration is common in tests of brittle materials for strength and crack growth resistance. No more appliances are required aside from a machine or press. This technique is based on the theoretically substantiated statement [23] that the homogeneous (containing no cracks) cylinder compressed by uniform distribution along a generatrix of forces with intensity P undergoes uniform tensile stress $\sigma = P/(\pi R)$ in the plane of compression, where R is the specimen's radius.

Consequently, the crack located in the plane of compression is under the same conditions of uniaxial tension as the crack in Griffith's problem for plane strain (Fig. 6.3).

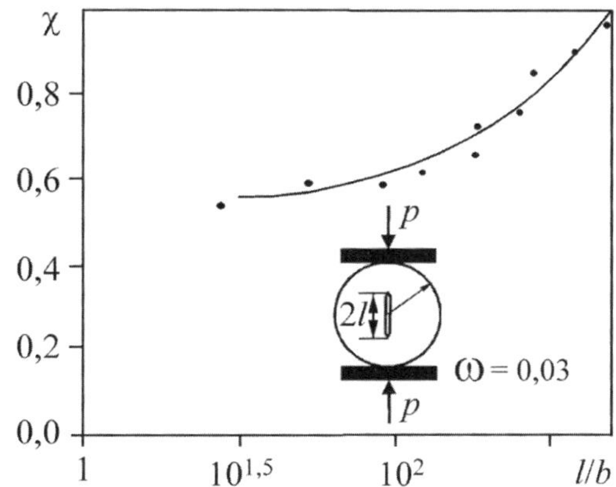

Fig. 6.10 The comparison of theoretical and experimental data: Solid line shows prediction using formula (6.44); points present experimental data for concrete specimens ($R = 50$ mm, $l = 0.25R$)

The disk with a crack loaded in the same configuration has SIF determined by the formula [24]:

$$K_{\mathrm{I}} = \frac{P}{R}\sqrt{\frac{l}{\pi}}\left[1 + \frac{3}{2}\lambda^2 + \frac{3}{4}\lambda^6 + O(\lambda^8)\right],$$

where $O(\lambda^8)$ is the small quantity of order of λ^8; $\lambda = l/R$.

It is easy to show that the free disk surface effect on K_{I} value is negligible (error less than 10 %) at $l/R \leq 0.25$. In such cases, K_{I} values can be approximately calculated using the simple formula

$$K_{\mathrm{I}} = \sigma\sqrt{\pi l}, \quad \sigma = \frac{P}{\pi R},$$

while SIF for the filled crack in the disk is determinable from Eq. (6.39).

Then, the formula (6.44) will give values of strengthening parameter χ for the cracked disk with the same error. In this way, we have everything necessary to compare theoretical predictions Eq. (6.44) and the experimental data related to the strengthening parameter $\chi = p_c/\sigma_B^*$.

The above presented data (Fig. 6.10) demonstrate that the above presented mathematical models for estimating serviceability (strength) of structural elements with healed cracks are in good accordance with experimental data and appropriate for implementation in engineering practice.

6.4 Strength Estimation for a Plate with a Filled Crack Using δ_C-criterion [25]

If the pre-failure zone size (Δl in Fig. 6.6) is comparable with the typical defect size (l_0 in Fig. 6.6), then the Griffith-Irwin concepts in application for determining the limit equilibrium state of a body with a crack need some refinements (the condition

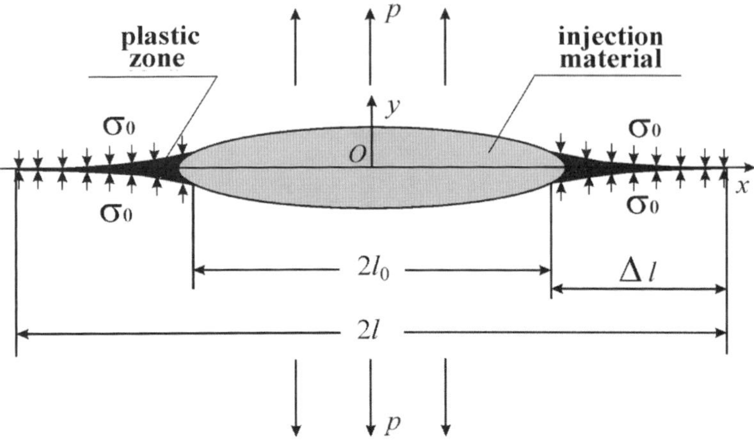

Fig. 6.11 The filled crack in the δ_c-model

$\Delta l \ll l_0$ is violated). In such cases, the δ_c-model is useful for determining the limit load ($p = p_*$) [16], [17] from the condition that $\delta_{p_*} = 2u_y(\pm l_0, 0, p_*) = \delta_{1C}$. Taking into consideration the expression (6.12) for crack edge displacement near the crack tip $2u_y(p, \pm l_0, 0) = \delta_p = -8l_0\sigma_0 c \ln \cos(\pi p/2\sigma_0)$, one can derive the following formula for the limit load:

$$p_* = \frac{2}{\pi}\sigma_0 \arccos(e^{-\eta}), \quad \eta = \frac{\delta_{1C}}{8cl_0\sigma_0}. \tag{6.45}$$

Let us now consider the case of a plate with a filled crack under uniaxial tension in terms of δ_c-model (Fig. 6.11).

Let elastic properties of the solidified filler be described by the Young modulus $E_1 < E$ and the Poisson ratio ν_1. Assume ideal mechanical adhesion at the material interface. The response of the filler material to the tensile forces in the body is representable using Eqs. (6.34) based on the model of the Winkler foundation.

As above, one can represent the problem in Fig. 6.11 as a superposition of two configurations, namely, the tension of the intact plate by a specific force p and tension of the plate with a crack loaded by the same specific force p. Since the first problem is trivial, the second problem obviously will determine the limit equilibrium state in the general problem.

In the second case, the elastic theory plane problem for tension of an infinite plate with a filled crack (Fig. 6.11) is reducible in frames of δ_c-model to the following boundary problem. The plate has a straight slit $-l \le x \le l$ with the following boundary conditions at its contour:

$$\begin{cases} \sigma_{yy}(x, 0) = -p + \dfrac{[u^*(x)]}{h(x)}E_1, 0 \le |x| \le l_0; \\ \sigma_{yy}(x, 0) = \sigma_0 - p, l_0 \le |x| \le l; \\ \sigma_{xy} = 0, 0 \le |x| \le l. \end{cases} \tag{6.46}$$

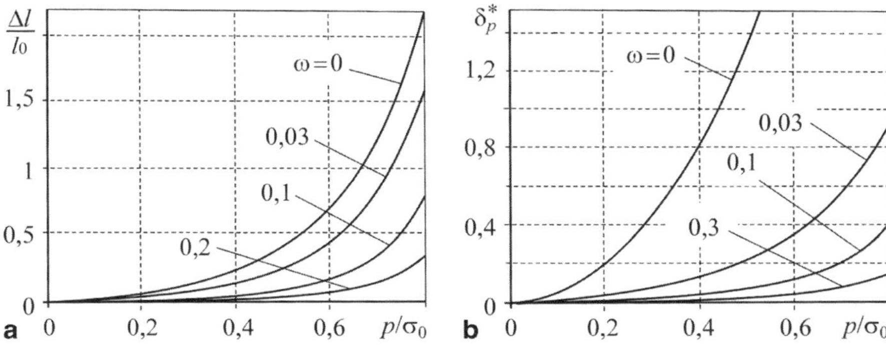

Fig. 6.12 Dependences of the plastic zone size (**a**) and rated crack opening (**b**) on load magnitude (p/σ_0) at different values of relative hardness $(\omega = E_1/E)$ of the filler material. The case of crack geometry $\beta = l_0/b = 10$

Stresses in infinitely remote points are absent:

$$\sigma_{yy}^\infty = \sigma_{xx}^\infty = \sigma_{xy}^\infty = 0.$$

Again, solving the elastic theory problem with conditions (6.46) is reducible to the singular integro-differential equation [25]:

$$\int_{-l}^{l} \frac{u_y'(t)}{t - x} dt - 2\pi c\omega \frac{u_y(x)}{d(x)} \cdot H(l_0 - |x|) = \frac{2\pi c}{E}(-p + p\omega \cdot H(l_0 - |x|) + \sigma_0$$
$$\cdot H(l_0 - |x|)),$$
$$|x| \leq l$$

$$(6.47)$$

Here $H(\zeta)$ is the Heaviside function $(\zeta = l_0 - |x|, \zeta = |x| - l_0)$, $\omega = E_1/E$. In this equation, u_y^0 are displacements of surface points in the intact (containing no cracks) body under applied forces p. Note that the plastic zone size $l - l_0$ near the filled crack tip in the δ_c-model is determined from stress restrictions in the points $x = \pm l$.

Some Partial Numerical Solutions

Known solutions of the singular integro-differential equation (6.47) use the Chebyshev–Gauss quadratures (see, e.g., [22]). Figures 6.12, 6.13 and 6.14 show curves illustrating dependences of the plastic zone size, the crack opening in the points $x = \pm l$, and ultimate load p_* on relative hardness (E_1/E) of the filler material and crack geometry $\beta = l_0/b$. Presented calculations assumed the initial crack shape as ellipsoidal with semi-axes l_0 and b $(l_0 \gg b)$.

The numerical solutions reveal that the hardness parameter E_1/E exerts a great impact on the pre-failure zone size, the crack opening between the edges near the crack tip, and, consequently, the ultimate load and strength of the body.

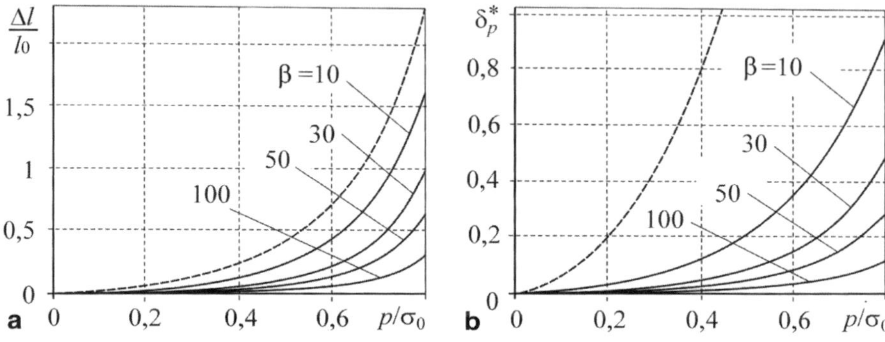

Fig. 6.13 Dependences of the plastic zone size (**a**) and rated crack opening (**b**) on load magnitude (p/σ_0) at different values of crack eccentricity ($\beta = l_0/b$). Relative filler hardness $E_1/E = 0.03$

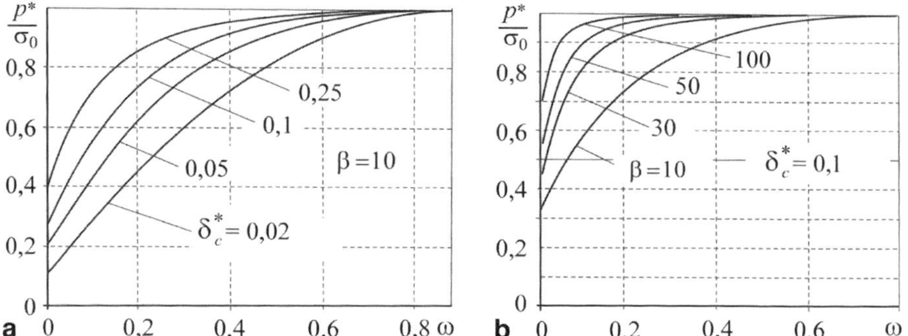

Fig. 6.14 Dependence of the ultimate load on relative hardness ($\omega = E_1/E$) of the filler material (**a**) and crack geometry $\beta = l_0/b$ (**b**)

Therefore, providing the adequate filler hardness (value ω) (Fig. 6.14b), one can restore the serviceability of the impaired material to a significant extent.

Another parameter influencing the efficiency of crack healing is the initial crack opening (or crack geometry). Cracks with lower initial crack opening are better healed (Fig. 6.15). They are liable to complete healing through a material with hardness by orders of magnitude less than the hardness of the host material. This feature of defect healing has an especially practical importance because the hardness of injection materials is, as a rule, much less than the hardness of structural materials.

Approximate Solution of the Problem

If we assume the opened crack elliptical in the cross-section with semi-axes l_0 and b, then stresses in the filler material will be approximately homogeneous or constant (in the absence of plastic zones inside the inclusion). Such an assumption allows for

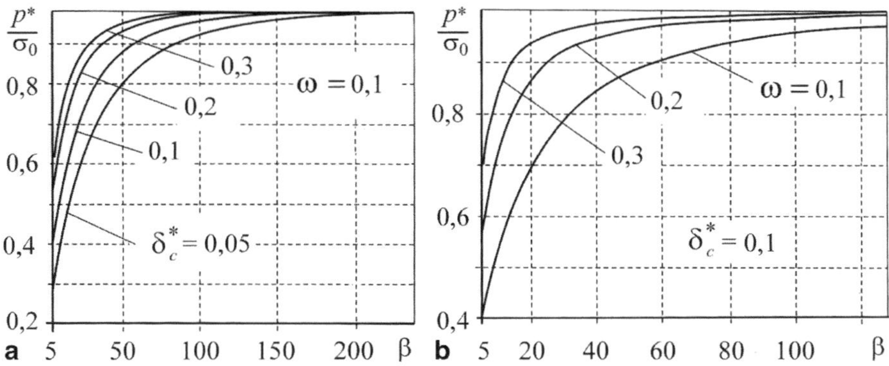

Fig. 6.15 Dependence of the ultimate load on crack geometry β at different values of displacement near the crack (**a**) or relative hardness ω of the filler material (**b**)

finding the exact solution of the singular integro-differential equation (6.47) [25]:

$$u_y(x) = \frac{c}{E}\left\{ 2\pi p(1 - \lambda k)\sqrt{l^2 - x^2} + (\sigma_0 - \lambda k p)\cdot \right. \tag{6.48}$$

$$\left. \left[(x - l_0)\Gamma(l, x, l_0) - (x + l_0)\Gamma(l, x, l_0) - 4\sqrt{l^2 - x^2}\arccos\frac{l_0}{l} \right] \right\}, 0 \le |x| \le l$$

where

$$k = \frac{1 + 2\beta}{1 + 2\beta\lambda}; \quad \beta = \frac{b}{l_0}; \quad \Gamma(l, x, l_0) = \ln\frac{l^2 - xl_0 - \sqrt{(l^2 - x^2)(l^2 - l_0^2)}}{l^2 - xl_0 + \sqrt{(l^2 - x^2)(l^2 - l_0^2)}}.$$

The plastic zone size $l - l_0$ in the axis x direction is once again determined from stress restriction and continuity in the points $x = \pm l$:

$$l - l_0 = l_0\left(\sec\frac{\pi p(1 - \lambda k)}{2(\sigma_0 - \lambda k p)} \right). \tag{6.49}$$

Solution (6.48), combined with relationship (6.49), yields displacements near the filled crack tip in the form:

$$u_y(l_0) = -\frac{4l_0 c}{E}(\sigma_0 - \lambda k p)\ln\cos\left(\frac{\pi p(1 - \lambda k)}{2(\sigma_0 - \lambda k p)} \right). \tag{6.50}$$

Now, the ultimate load p_*, accounting for the critical crack-opening criterion $2u_y(p_*, l_0) = \delta_{1C}$, becomes:

$$\delta_{1C} = -\frac{8l_0 c}{E}(\sigma_0 - \lambda k p_*)\ln\cos\left(\frac{\pi p(1 - \lambda k)}{2(\sigma_0 - \lambda k p_*)} \right). \tag{6.51}$$

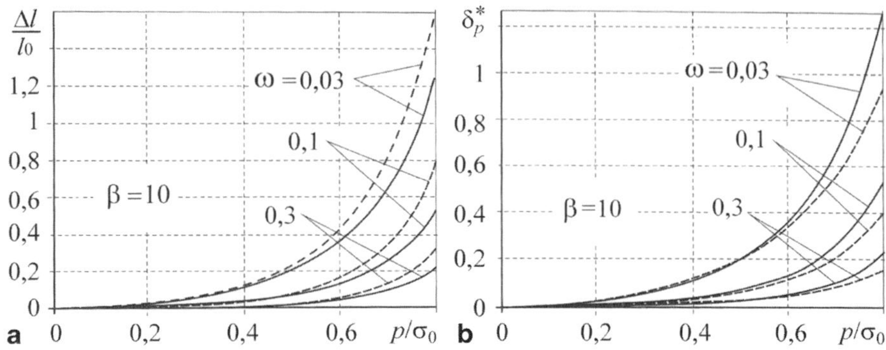

Fig. 6.16 Dependence of the plastic zone size (**a**) and crack opening (**b**) on load magnitude (p/σ_0) at different values of filler hardness $\omega = E_1/E$

Let us compare (Fig. 6.16) the numerical solutions of Eq. (6.47) obtained using the Chebyshev–Gauss quadrature (dashed lines) and approximate solution in the assumption of homogeneous stresses in inclusion (6.48) and (6.49) (solid lines). The comparison shows (Fig. 6.16) that they are in good agreement. Such a result enables us to believe that the approximate approach is a good enough method for estimating the structure renewal efficiency in engineering practice, as well as optimizing the material selection for injection technologies.

6.5   Strengthening of Damaged Materials at Partial Filling of Defect [26]

We have investigated above the material renewal efficiency using injection technologies under the assumption of a defect/crack entirely filled by injection material, which is rarely implementable in practice. In this section, we shall analyze the possible cases of partial filling of the crack with length $2l$ by an injection material and its influence on the efficiency of material strengthening.

1. Let us first consider the case when the injection material fills the crack beginning from its center (Fig. 6.17), in the segment $2l_1$.

After curing of the injection material, a tensile load begins to act on the body with intensity p. The integral equation describing such an injection and loading configuration has the form [26]:

$$\int_{-l}^{l} \frac{u_y'(t)dt}{t-x} - \frac{2\pi(1-v^2)E_1 u_y(x)}{Eh(x)} H(l_1-|x|) = \frac{2\pi(1-v^2)}{E} p\left(\frac{E_1}{E}H(l_1-|x|)-1\right),$$

$$|x| \leq l.$$

$$(6.52)$$

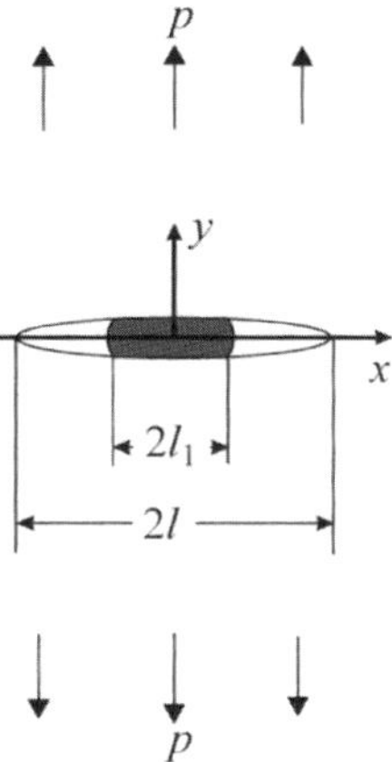

Fig. 6.17 Partially filled crack in an infinite body

Numerical solutions of Eq. (6.52) yield curves as presented in Fig. 6.18 illustrating stress relief in the crack vicinity depending on the injection material's filling degree (l_1/l), the crack geometry, and the hardness of the filler. As seen from the figure, all mentioned factors influence the efficiency of strengthening. These calculations correspond to the elliptical crack surface shape, i.e. $h(x) = b\sqrt{1 - x^2/l^2}$.

In Fig. 6.18a, solid lines correspond to filling of the crack by the injection material with hardness $\omega = 0.03$, which is typical for the system of concrete-polyurethane (concrete being the host material, polyurethane the foreign injection material). Dashed lines describe the harder injection material with hardness of $\omega = 0.1$.

Figure 6.18 demonstrates a dramatic decrease of the stress intensity factors and, hence, the strengthening effect when the crack approaches complete filling $l_1 \rightarrow l$.

At partial filling (Fig. 6.17), essential strengthening of the body takes place only at a high enough filling degree. More specifically, a crack filling degree as high as about 80 % ensures only 50 % of highest possible strengthening for the given injection material.

2. Now let us fill the crack with an injection material only at the crack tips as shown in Fig. 6.19. Then, the respective integral equation for such a configuration has the form [26]:

$$\int_{-l}^{l} \frac{u'_y(t)dt}{t - x} - \frac{2\pi(1 - v^2)E_1 u_y(x)}{Eh(x)} H(|x| - l_1) = \frac{2\pi(1 - v^2)p}{E} \left(\frac{E_1}{E} H(|x| - l_1) - 1 \right),$$

$$|x| \leq l.$$

$$(6.53)$$

As above, we obtained the illustrative numerical solution in the Chebyshev–Gauss mechanical quadrature assuming the crack surface to be an elliptical cylinder with semi-axes $b \ll l$. The results of numerical analysis (Fig. 6.20) indicate that strengthening of a body is especially sensitive to injection into the crack tips. In particular, by filling the crack tips along only 20 % of the crack length, the percentage of body

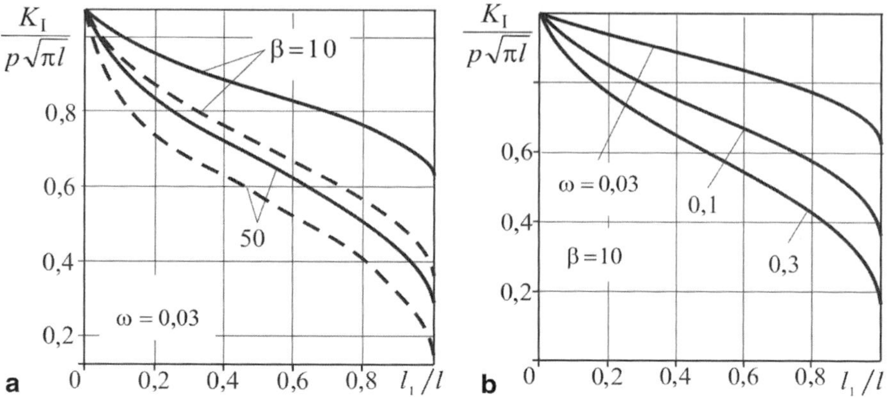

Fig. 6.18 Dependence of SIF near the crack tip on the filling zone size (l_1/l) at various values of the crack geometry factor $\beta = l/b$ (**a**) and filler hardness $\omega = E_1/E$ (**b**)

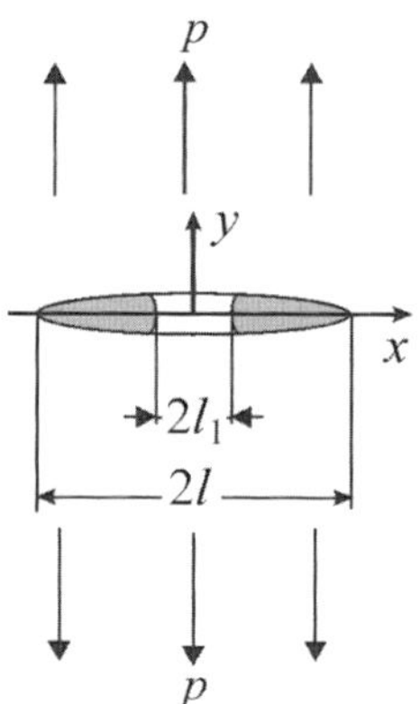

Fig. 6.19 Body with crack partially filled by injection material in tips

strengthening reaches 85 % of the highest possible degree for the given injection material.

3. Finally, let us consider the crack injected as shown in Fig. 6.21.

The respective singular integro-differential equation has the form [26]:

$$\int_0^{2l} \frac{u_y'(t)dt}{t - x} - \frac{2\pi(1 - v^2)E_1 u_y(x)}{Eh(x)} H(l_1 - x) = \frac{2\pi(1 - v^2)p}{E} \left(\frac{E_1}{E} H(l_1 - x) - 1 \right),$$
$$0 < x < 2l.$$

$$(6.54)$$

The numerical solution of the problem (Fig. 6.22) illustrates SIF changes near each crack tip at various values of filler hardness and crack filling degree. The most effective is strengthening of the starting crack tip whereas the opposite tip remains most dangerous. The SIF near the finishing (right) crack tip decreases only at $l_1 \to 2l$.

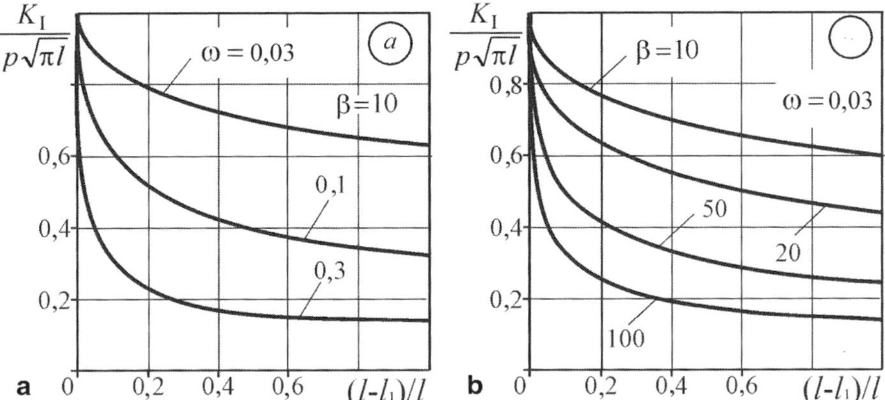

Fig. 6.20 Dependence of SIF near the crack tip on the filling degree $(l - l_1)/l$ at various values of crack geometry factor $\beta = l/b$ (**a**) and filler hardness $\omega = E_1/E$ (**b**)

Fig. 6.21 Body with crack partially filled by injection material from one side

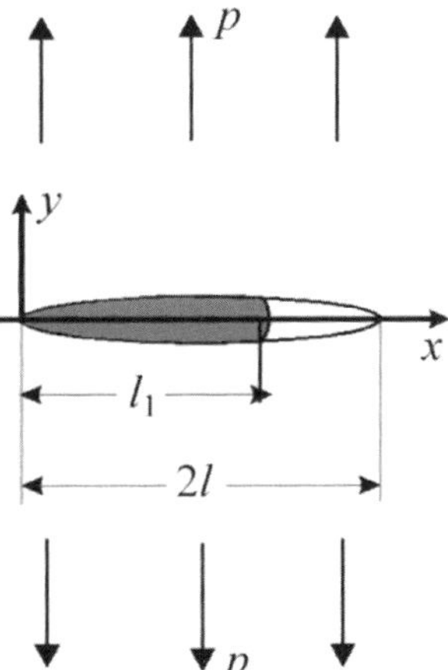

6.6   Crack Wedging Effect During Injection

The defect of the crack wedging by injection mixture is an undesirable, harmful effect (from the point of view of structural integrity) caused by the injecting pressure that is inevitable for implementation of the given process. The injection mixture presses onto the surface of the defect or crack initially while arriving at the crack and then during solidification. Combined with applied loads, such wedging can induce crack growth. Let us analyze this phenomenon in terms of fracture mechanics.

Let an injection material possessing the elastic characteristics E_1 and ν_1 after solidification fill a defect loaded by applied force intensity p_0. The material exerts pressure onto the crack surface of an unknown, generally speaking, magnitude $\tilde{p}$, which is the sum of two components: (a) hydrostatic pressure p_1 transmitted by the

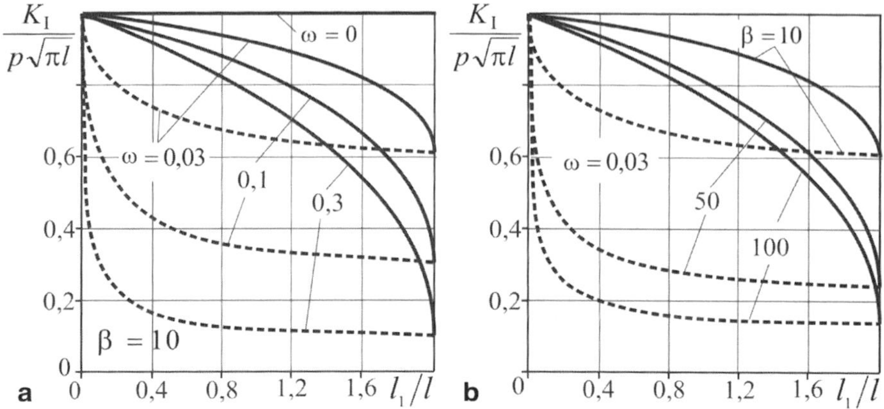

Fig. 6.22 Dependence of SIF near the crack tips on the filling zone size (l_1/l) at various values of filler hardness $\omega = E_1/E$ (**a**) and inclusion shape factor $\beta = l/b$ (**b**). *Dashed lines* correspond to the filled (*left*) crack tip; *solid lines* to the empty (*right*) tip.

liquid injection material, and (b) pressure p_2 arising due to material volume change during solidification.

Therefore, total pressure acting on the surface of the defect or crack is as follows:

$$\tilde{p} = p_1 + p_2.$$

The first summand p_1 being known, the second one is subject to determination by solving the respective problem of fracture mechanics. Suppose that the liquid injection material with volume V_0 in absence of a counteraction of the host material changes volume during solidification into $V_1 = \acute{\alpha} V_0$. The counteraction of crack surfaces and related deformations results in another volume V_2 of deformed injection material limited by the interface of the materials.

An approximate expression for stresses in the injection material arising as the result of resistance of the host material has the form:

$$p_2 = E_1 \left(\frac{u_0 + u_1 + u_2}{h} - 1 \right). \tag{6.55}$$

Here, u_0, u_1, and u_2 are displacements of the defect/crack edges caused by forces p_0, p_1, and p_2, respectively; $h = b\sqrt{1 - x^2/l^2}$; $b = V_1/\pi l t = \alpha l(p_0 + p_1)/E$;t is the plate's thickness.

Given the displacement of the crack edges u_0 and u_1 in the form:

$$u_0 = \frac{p_0}{E}\sqrt{l^2 - x^2},$$

$$u_1 = \frac{p_1}{E}\sqrt{l^2 - x^2}, \tag{6.56}$$

we can determine u_2 by solving the integral equation:

$$\int_{-l}^{l} \frac{u'_y}{t-x} dt - \frac{2\pi E_1(1-v)u_2}{E(1+v)h} = \frac{2\pi(1-v)E_1}{E(1+v)}\left(\frac{(p_0+p_1)\beta}{E} - 1\right), \qquad (6.57)$$

which has an exact solution given by:

$$u_2 = \frac{\omega(E-(p_0+p_1)\beta)}{E(1+\beta\omega)}\sqrt{l^2-x^2}, \omega \equiv \frac{E_1}{E}. \qquad (6.58)$$

From here and from known solutions of the problem regarding a plate with a straight crack under tension, we can derive the relationship for total SIF from cumulative effects of all factors for the filled crack:

$$K_{\mathrm{I}} = \sqrt{\pi l}\frac{p_0+p_1+\omega(E-(p_0+p_1)\beta)}{1+\beta\omega}. \qquad (6.59)$$

Now, we can express the highest permissible injection mixture feed pressure p_1 corresponding to the crack start from the condition $K_{\mathrm{I}} < K_{\mathrm{IC}}$:

$$p_1 < \frac{K_{\mathrm{IC}}(1+\beta\omega)}{\sqrt{\pi l}} - p_0 - \omega E. \qquad (6.60)$$

Given fixed values of both applied load p_0 and processing mixture feed pressure p_1, the inequality $K_{\mathrm{I}} \leq K_{\mathrm{IC}}$ yields the solid injection material hardness corresponding to the crack start:

$$\omega \leq \frac{(p_0+p_1)\sqrt{\pi l} - K_{\mathrm{IC}}}{E\sqrt{\pi l} - \beta K_{\mathrm{IC}}}. \qquad (6.61)$$

It should be noted that at pressure values p_1 and p_2, low when compared with σ_B or p^* for host material with an empty crack, these factors can cause only local slow crack growth, i.e., the wedging effect is negligible.

6.7 Three-Dimensional Problems of Strengthening a Body with a Crack

Let us consider solutions of selected spatial (three-dimensional, 3D) problems related to filled cracks in a body. General solutions of such problems entail great mathematical difficulties. Therefore, we shall confine this study to canonical crack configurations (circular or elliptical cracks) with the aim of forming conclusions about the spatial effects on the strength of bodies with filled cracks and the validity range of the two-dimensional models of deformable solid bodies with cracks.

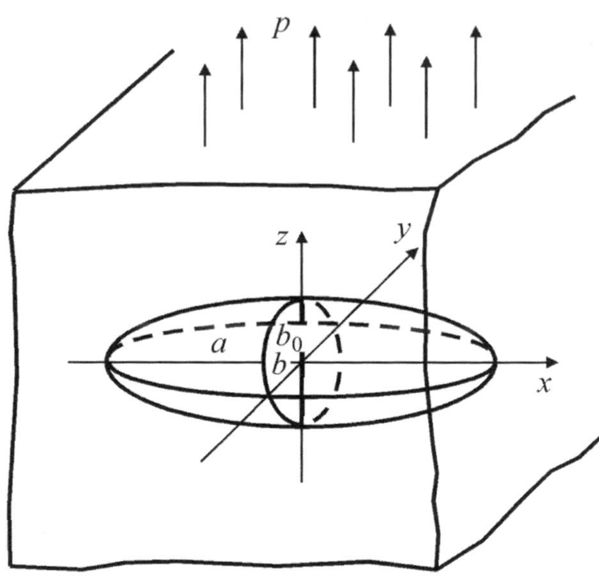

Fig. 6.23 Tension of an infinite elastic body with an elliptical crack

Strength of Bodies with Elliptical Cracks Under Tension

Let a body with an elliptical in front view crack-like defect occur under uniaxial tension by forces with intensity p applied perpendicularly to the crack plane (Fig. 6.23).

SIF for the infinite body with such a crack has the form [4]:

$$K_{\mathrm{I}} = \frac{p\sqrt{\pi b}(a^2\sin^2\varphi + b^2\cos^2\varphi)^{1/4}}{E(k)\sqrt{a}}, \tag{6.62}$$

where a, b are ellipse semi-axes; ϑ is one of the polar coordinates determining a point position on the ellipse $x^2/a^2 + y^2/b^2 = 1$ (Fig. 6.24), and $E(k)$ is a complete elliptic integral of the second kind:

$$E(k) = \int_0^{\pi/2} \sqrt{1 - k\sin^2\alpha}\,d\alpha, k \equiv \frac{a^2 - b^2}{a^2}. \tag{6.63}$$

The strength of such a cracked body results from the fracture criterion (6.1) that has the following appearance for the given case:

$$K_{\mathrm{I}}^* = K_{\mathrm{IC}}, \tag{6.64}$$

where K_{I}^* is a maximum value of the intensity factor $K_{\mathrm{I}}(\varphi) = K_{\mathrm{I}}(\pi/2\,)$; K_{IC} is the fracture toughness of the material.

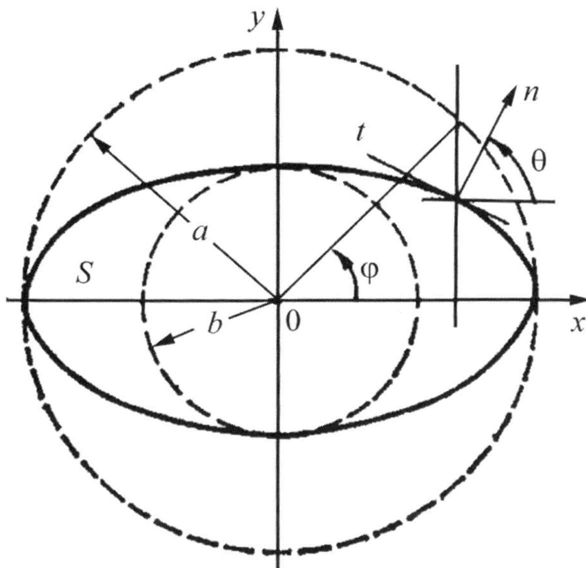

Fig. 6.24 The median area S of an elliptical crack and the related coordinate systems

Combining Eqs. (6.62) and (6.64), one can easily obtain relationships [4] for the fracture stress in a bulk body with an inner elliptical crack:

$$p_c = \frac{K_{IC} E(k)}{\sqrt{\pi b}}.$$

(6.65)

This formula is valid for a body with an empty elliptical crack.

Let us now consider the crack-like defect filled by injection material in a whole volume. Then, we have a system of integral equations (6.31), which reduces for the above problem to the single equation [6]:

$$\nabla^2_{xy} \iint\limits_S \frac{u_z(\xi,\eta)d\xi\,d\eta}{\sqrt{(x-\xi)^2 + (y-\eta)^2}} - \frac{4\pi(1-v^2)}{E}\frac{u_z(x,y)}{h(x,y)} = \frac{4\pi(1-v^2)p(\omega-1)}{E}, \quad \omega = \frac{E_1}{E}.$$

(6.66)

Its solution lies in the assumption that the surface S of the ellipse has the form [6]:

$$u_z(x,y) = \frac{2b_0 p(1-v^2)\beta(1-\omega)}{E[2\beta\omega(1-v^2) + E(k)]}\sqrt{1 - \frac{x^2}{a^2} - \frac{y^2}{b^2}}$$

(6.67)

where $\beta = b/b_0$ and b_0 is the shortest of the ellipse semi-axes.

Using expressions (6.32), one can derive SIF for the filled crack. To do so, let us first change from rectangular Cartesian coordinates (x, y, z) to parametrical coordinates (t, n, z) with origin at the crack contour (Fig. 6.24):

$$x = a \cos \varphi - t \sin \theta + n \cos \theta,$$

$$y = b \sin \varphi + t \cos \theta + n \sin \theta, \tag{6.68}$$

$$z = z$$

The angle θ is expressible through the parametrical equations of the ellipse by the relationships:

$$\cos \theta = \frac{b \cos \varphi}{\sqrt{a^2 \sin^2 \varphi + b^2 \cos^2 \varphi}},$$
$$\sin \theta = \frac{a \sin \varphi}{\sqrt{a^2 \sin^2 \varphi + b^2 \cos^2 \varphi}}. \tag{6.69}$$

Then, the asymptotic expression for displacements in close vicinity of the elliptical crack contour takes the form:

$$u_z = \frac{2\sqrt{2} b_0 p (1 - v^2)(1 - \omega)\beta(a^2 \sin^2 \varphi + b^2 \cos^2 \varphi)^{1/4}}{E[2\beta\omega(1 - v^2) + E(k)]\sqrt{ab}} \sqrt{-n} + O(n) \tag{6.70}$$

and intensity factor K_{I} of the filled crack in accordance with the first from relationships (6.32) becomes:

$$K_{\mathrm{I}} = \frac{\sqrt{\pi b} p (1 - \omega)\beta(a^2 \sin^2 \varphi + b^2 \cos^2 \varphi)^{1/4}}{[2\beta\omega(1 - v^2) + E(k)]\sqrt{a}}. \tag{6.71}$$

From here and Eq. (6.5), we can calculate the fracture stress for the body with a filled crack as

$$p_c = \frac{\tilde{K}_{\mathrm{IC}}[2\beta\omega(1 - v^2) + E(k)]}{\sqrt{\pi b}(1 - \omega)}. \tag{6.72}$$

The comparison of ultimate loads expressed by Eqs. (6.65) and (6.72) enables us to establish strengthening efficiency at the application of the injection technology to a body with the crack.

To do this, we have to know two constants: fracture toughness K_{IC} of the material and its analogue in the case of filled crack $\tilde{K}_{\mathrm{IC}}$. In general, $\tilde{K}_{\mathrm{IC}}$ value can differ from K_{IC}. Experimental methods for determination of K_{IC} are well known and quite simple. However, no similar methods for determination of $\tilde{K}_{\mathrm{IC}}$ present in the literature at present. There we encounter only assumptions [27] that $\tilde{K}_{\mathrm{IC}}$ values are perhaps somewhat higher than the K_{IC} values. The present work presents for the first time the mathematical grounding for such a method and experimental data on this parameter.

In particular, we have established experimentally for concrete as the host material and polyurethane as the filler that values of $\tilde{K}_{\mathrm{IC}}$ and K_{IC} are very close one to another. We believe that this conclusion is also valid for other pairs of material where hardness of the filler is smaller than the hardness of the matrix.

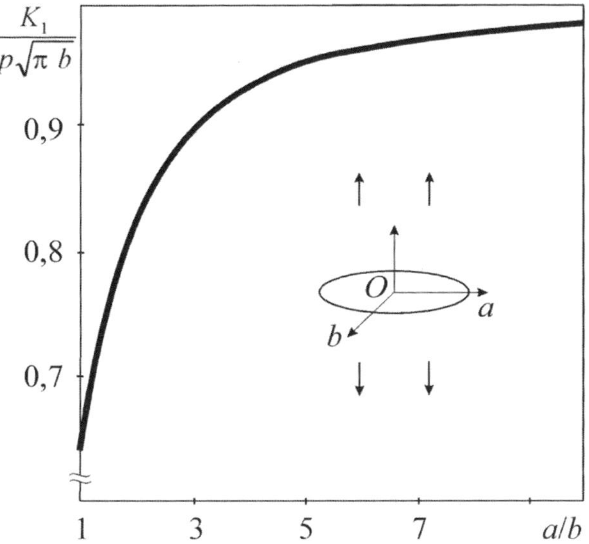

Fig. 6.25 Dependence of maximum SIF values at the elliptical crack front on the ellipse semi-axes ratio a/b. The case of an empty crack

Let σ_B be the ultimate strength of material subject to determination in smooth specimens, as recommended.

Then, the ratio $p_c/\sigma_B = \chi$ will characterize the degree of load-carrying capability renewal for the bulk body with an elliptical crack in the present 3D configuration as well.

$$\chi = \frac{K_{IC}^i [2\beta\omega(1 - v^2) + E(k)]}{\sigma_B \sqrt{\pi b}(1 - \omega)}. \tag{6.73}$$

It follows from the above dependence, as well as physical considerations, that the degree of load-carrying capability renewal (χ) varies from $K_{IC} E(k)/(\sigma_B \sqrt{\pi b})$ to 1. The upper limit $\chi = 1$ means a completely healed crack and restored strength while the lower limit $\chi = K_{IC} E(k)/(\sigma_B \sqrt{\pi b})$ corresponds to a non-healed (empty) crack.

One can enhance the strength renewal effect by either elevating the parameter β, i.e., by injecting at the narrowest possible crack opening, or increasing the parameter ω that characterizes the filler's hardness.

In order to illustrate the effect of spatial defect configuration on the strength of a body with an elliptical crack, Fig. 6.25 presents the dependences of SIF (K_1) in points of its maximum values ($\varphi = \pi/2$ according to Eq. (6.62)) on the a/b ratio. When the a/b ratio reaches high values, in other words, when the elliptical crack elongates up to the narrow strip (the problem becomes two-dimensional), the stress state near the crack contour approaches the two-dimensional Griffith model with crack length $2b$ under plane strain condition.

It follows from the above plot that solutions for the spatial cracks with semi-axes ratio $a/b > 5$ are virtually identical to solutions for plane problems. The difference in SIF at $a/b = 5$ is below 5 %. However, one should only use spatial solutions when

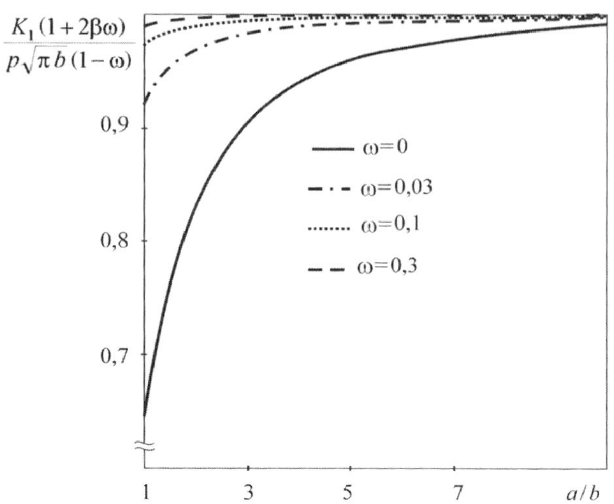

Fig. 6.26 Dependence of maximum SIF values at the elliptical crack front on the ellipse semi-axes ratio a/b. The case of a filled crack

semi-axes a and b are comparable. Figure 6.26 demonstrates similar dependences for filled cracks plotted using Eq. (6.71).

The asymptote in the plot corresponds to Griffith's crack filled by injection material and characterized by SIF value

$$K_{\mathrm{I}} = \frac{p\sqrt{\pi b}(1 - \omega)}{1 + 2\beta\omega}.$$

Strength Renewal for a Body with an Elliptical Crack Under Shear

Let us consider here the elliptical crack in an infinite elastic body under shearing loading (Fig. 6.27). SIF for such a configuration is subject to calculation using the following formulae [5]:

$$
\begin{aligned}
K_{\mathrm{II}} &= \frac{\sqrt{\pi q}\, bk^2}{\sqrt{ab}\left(a^2\sin^2\varphi + b^2\cos^2\varphi\right)^{1/4}} \\
&\quad \left(\frac{b\cos\alpha\cos\varphi}{E(k)(k^2 - v) + cvk'^2 K(k)} + \frac{a\sin\alpha\sin\varphi}{E(k)(k^2 - vk'^2) + vk'^2 K(k)}\right); \\
K_{\mathrm{III}} &= \frac{\sqrt{\pi q}(1 - v)bk^2}{\sqrt{ab}\left(a^2\sin^2\varphi + b^2\cos^2\varphi\right)^{1/4}} \\
&\quad \left(\frac{b\cos\alpha\sin\varphi}{E(k)(k^2 - v) + vk'^2 K(k)} + \frac{b\sin\alpha\cos\varphi}{E(k)(k^2 + vk'^2) - vk'^2 K(k)}\right).
\end{aligned}
\tag{6.74}
$$

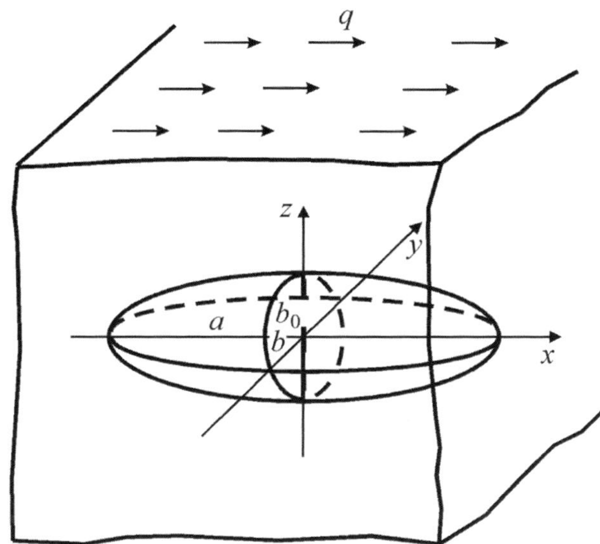

Fig. 6.27 Shearing of an infinite elastic body with an elliptical crack

Here $K(k) \equiv \int_0^{\pi/2} d\alpha / \sqrt{1 - k^2 \sin^2 \alpha}$ is a complete elliptic integral of the first kind; $k' = b/a$; α is the angle between shear direction and the longer semi-axis of ellipse $x^2/a^2 + y^2/b^2 - 1 = 0$.

Relationships (6.74) indicate the simultaneous existence of both crack edge shearing modes and transverse and longitudinal shears in the present case (see Fig. 6.4). They imply that the stress state near the crack contour is a combination of SIFs K_{II} and K_{III}. Unlike the foregoing, the contour points corresponding to the crack start are here unknown in advance because the stress component or component combination which will first attain the extreme value is unknown.

However, we can specify the contour points in advance, where either pure transverse shear or pure longitudinal shear arises. If the shear stress line q is directed perpendicularly to the shorter ellipse semi-axis, then in points $\varphi = 0$

$$K_{II} = \frac{\sqrt{\pi} b k^2 q}{\sqrt{a}[E(k)(k^2 - \nu) + \nu k'^2 K(k)]}, \qquad K_{III} = 0. \tag{6.75}$$

Conversely, in points with $\varphi = \pi/2$:

$$K_{II} = 0, \qquad K_{III} = \frac{\sqrt{\pi}(1 - \nu)\sqrt{b} k^2 q}{E(k)(k^2 - \nu) + \nu k'^2 K(k)}. \tag{6.76}$$

The bulk body with an elliptical crack under shearing in planes parallel to the crack plane will fracture in transverse shear mode if the condition

$$K_{II} = K_{IIc} \tag{6.77}$$

will be attained first, where K_{IIC} is the ultimate value of SIF K_{II}.

In such a case, the fracture load q_C can be written from Eqs. (6.75) and (6.77) as:

$$q_C = \frac{K_{\mathrm{IIC}}\sqrt{a}[E(k)(k^2 - v) + vk'^2 K(k)]}{\sqrt{\pi}\,bk^2}.$$

(6.78)

On the other hand, if the condition

$$K_{\mathrm{III}} = K_{\mathrm{IIIc}},$$

(6.79)

will be attained in points $\varphi = \pi/2$, then the body will fracture in longitudinal shear mode, where K_{IIIC} is the constant describing the resistance of the material to crack propagation in such a mode.

The fracture load q_C here will have the form:

$$q_C = -\frac{K_{\mathrm{IIIC}}\sqrt{a}[E(k)(k^2 - v) + vk'^2 K(k)]}{\sqrt{\pi}\,b(1 - v)k^2}.$$

(6.80)

Let us fill the crack with an injection material. Then, the crack edge deformations will be representable by the system of two equations as follows from Eq. (6.31) [6]:

$$
\begin{aligned}
u_x(x, y) - \frac{h}{2\pi(1 - v)\omega'} &\left[\left(\frac{\partial^2}{\partial x^2} + (1 - v)\frac{\partial^2}{\partial y^2}\right) \iint_S \frac{u_x(\xi, \eta)\,d\xi\,d\eta}{\sqrt{(x - \xi)^2 + (y - \eta)}}\right. \\
&\left. + \frac{\partial^2}{\partial x \partial y} \iint_S \frac{u_y(\xi, \eta)\,d\xi\,d\eta}{\sqrt{(x - \xi)^2 + (y - \eta)}}\right] = \frac{q \cos\alpha(1 - \omega')h}{\mu_1};
\end{aligned}
$$

$$
\begin{aligned}
u_y(x, y) - \frac{h}{2\pi(1 - v)\omega'} &\left[\left(\frac{\partial^2}{\partial y^2} + (1 - v)\frac{\partial^2}{\partial x^2}\right) \iint_S \frac{u_y(\xi, \eta)\,d\xi\,d\eta}{\sqrt{(x - \xi)^2 + (y - \eta)}}\right. \\
&\left. + \frac{\partial^2}{\partial x \partial y} \iint_S \frac{u_x(\xi, \eta)\,d\xi\,d\eta}{\sqrt{(x - \xi)^2 + (y - \eta)}}\right] = \frac{q \sin\alpha(1 - \omega')h}{\mu_1};
\end{aligned}
$$

$$\omega' \equiv \frac{\mu_1}{\mu}.$$

(6.81)

The continuous solution of the system (6.81) under assumption of an elliptical crack shape is as follows [6]:

$$u_x(x, y) = \frac{q(1 - v)(1 - \omega')bk^2 \cos\alpha}{\mu[b_0 E(k)(k^2 - v) + b_0 vk'^2 K(k) + bk^2(1 - v^2)\omega']}h;$$

$$u_y(x, y) = \frac{q(1 - v)(1 - \omega')bk^2 \sin\alpha}{\mu[b_0 E(k)(k^2 + vk'^2) - b_0 vk'^2 K(k) + bk^2(1 - v^2)\omega']}h.$$

(6.82)

Using formulae (6.68) and (6.69) to change to the local coordinate system (t, n, z), one can re-write Eq. (6.82) in the new coordinates as:

$$u_x = \frac{\sqrt{2}q\cos\alpha(1-v)(1-\omega')b_0bk^2(a^2\sin^2\varphi+b^2\cos^2\varphi)^{1/4}}{\mu[b_0E(k)(k^2-v)+b_0vk'^2K(k)+bk^2(1-v^2)\omega']\sqrt{ab}}\times\sqrt{-n}+O(n);$$

$$u_y = \frac{\sqrt{2}q\sin\alpha(1-v)(1-\omega')b_0bk^2(a^2\sin^2\varphi+b^2\cos^2\varphi)^{1/4}}{\mu[b_0E(k)(k^2+vk'^2)-b_0vk'^2K(k)+bk^2(1-v)\omega']\sqrt{ab}}\times\sqrt{-n}+O(n),$$

$$(6.83)$$

where terms of the order of $O(n)$ are negligible.

The conversion formulae for change from displacement components u_x, u_y in the coordinate system xyz to displacements u_t, u_n in local coordinate system tnz have the appearance:

$$u_t = -u_x\sin\theta+u_y\cos\theta,$$

$$u_n = u_x\cos\theta+u_y\sin\theta,$$

$$(6.84)$$

where angle θ is subject to determination from relationships (6.69). Combining all of the above formalism, we can express the components of local displacements as:

$$u_t = \left[\frac{q\cos\alpha(1-v)(1-\omega')b_0bk^2a\sin\varphi}{\mu[b_0E(k)(k^2-v)+b_0vk'^2K(k)+bk^2(1-v)\omega']}\right.$$

$$\left.+\frac{q\sin\alpha(1-v)(1-\omega')b_0b^2k^2\cos\varphi}{\mu[b_0E(k)(k^2+vk'^2)-b_0vk'^2K(k)+bk^2(1-v)\omega']}\right]$$

$$\times\frac{\sqrt{-2n}}{(a^2\sin^2\varphi+b^2\cos^2\varphi)^{1/4}\sqrt{ab}}+O(n);$$

$$u_n = \left[\frac{q\cos\alpha(1-v)(1-\omega')b_0b^2k^2a\cos\varphi}{\mu[b_0E(k)(k^2-v)+b_0vk'^2K(k)+bk^2(1-v)\omega']}\right.$$

$$\left.+\frac{q\sin\alpha(1-v)(1-\omega')b_0bak^2\sin\varphi}{\mu[b_0E(k)(k^2+vk'^2)-b_0vk'^2K(k)+bk^2(1-v)\omega']}\right]$$

$$\times\frac{\sqrt{-2n}}{(a^2\sin^2\varphi+b^2\cos^2\varphi)^{1/4}\sqrt{ab}}+O(n).$$

$$(6.85)$$

Now, based on the last two formulas (6.32), we can write the intensity factors for the filled crack:

$$K_{\mathrm{II}}^i = \frac{\sqrt{\pi}q(1-\omega')b_0bk^2}{\sqrt{ab}(a^2\sin^2\varphi+b^2\cos^2\varphi)^{1/4}}\left[\frac{b\cos\alpha\cos\varphi}{b_0E(k)(k^2-v)+b_0vk'^2K(k)+bk^2(1-v)\omega'}\right.$$

$$\left.+\frac{a\sin\alpha\sin\varphi}{b_0E(k)(k^2+vk'^2)-b_0vk'^2K(k)+bk^2(1-v)\omega'}\right];$$

$$K_{\text{III}}^i = \frac{\sqrt{\pi}q(1-v)(1-\omega')b_0bk^2}{\sqrt{ab}(a^2\sin^2\varphi + b^2\cos^2\varphi)^{1/4}} \left[\frac{a\cos\alpha\sin\varphi}{b_0E(k)(k^2-v) + b_0vk'^2K(k) + bk^2(1-v)\omega'} \right.$$

$$\left. + \frac{b\sin\alpha\cos\varphi}{b_0E(k)(k^2+vk'^2) - b_0vk'^2K(k) + bk^2(1-v)\omega'} \right]. \tag{6.86}$$

Applying criterion (6.2) to the above-derived expressions, we can write formulae for critical loads causing fracture of the body with a filled elliptical crack in the selected configurations. If the direction of shearing force q coincides with the longer ellipse axis, that is, $\alpha = 0$, then, in points $\varphi = 0$ the purely transverse shear of the crack edges will take place with SIF

$$K_{\text{II}}^i = \frac{\sqrt{\pi}b_0bk^2q(1-\omega')}{\sqrt{a}[b_0E(k)(k^2-v) + b_0vk'^2K(k) + bk^2(1-v)\omega']}; \tag{6.87}$$

$$K_{\text{III}}^i = 0.$$

For this specific configuration, from Eq. (6.87) and criterion $K_{\text{II}}^i = K_{\text{IIC}}^i$, the fracturing load is

$$q_c = \frac{K_{\text{IIC}}^i\sqrt{a}\left[b_0E(k)\left(k^2-v\right) + b_0vk'^2K(k) + bk^2(1-v)\omega'\right]}{\sqrt{\pi}b_0bk^2(1-\omega')} \tag{6.88}$$

The degree of load-carrying capability renewal for the body with an elliptical crack is defined here as $\chi = q_C/\tau_C$, where τ_C is the ultimate shear strength, and has the value

$$\chi_{\text{II}} = \frac{K_{\text{IIC}}^i\sqrt{a}\left[b_0E(k)\left(k^2-v\right) + b_0vk'^2K(k) + bk^2(1-v)\omega'\right]}{\tau_c\sqrt{\pi}b_0bk^2(1-\omega')} \tag{6.89}$$

It follows from this expression that the essential parameters influencing the renewal efficiency for a body with an elliptical crack are, again, the filler's hardness factor (ω') and crack geometry (a, b, b_0).

Similarly, in points $\varphi = \pi/2$ the purely longitudinal shear will take place. We obtain from Eqs. (6.86) for this case:

$$K_{\text{III}}^i = \frac{\sqrt{\pi}bb_0k^2(1-v)(1-\omega')q}{b_0E(k)(k^2-v) + b_0vk'^2K(k) + bk^2(1-v)\omega'} \tag{6.90}$$

The degree of strength renewal for the structural element with elliptical crack χ_{III} after filling the crack with injection material will be as follows:

$$\chi_{\text{III}} = \frac{K_{\text{IIIC}}^i\left[b_0E(k)\left(k^2-v\right) + b_0vk'^2K(k) + bk^2(1-v)\omega'\right]}{\tau_c\sqrt{\pi}bb_0k^2(1-v)(1-\omega')}. \tag{6.91}$$

The same conclusion follows from Eq. (6.91) that the healing efficiency depends on the filler's hardness and crack geometry just as in the case of the transverse shear.

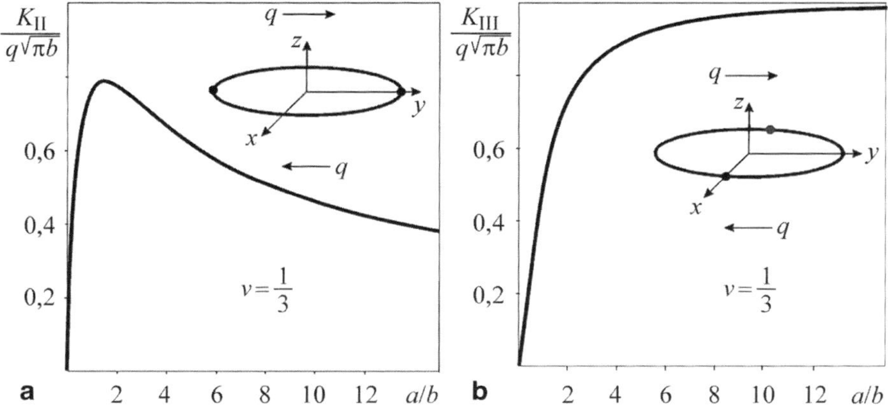

Fig. 6.28 SIF dependence on parameter a/b under transverse (**a**) or longitudinal (**b**) shear

Let us compare solutions for the strength of bodies with filled or empty cracks under shear loading obtained in two-dimensional and three-dimensional (spatial) approaches. Note that the straight Griffith crack with length $2b$ under transverse or longitudinal shearing has SIFs, respectively [22]:

$$K_{\mathrm{II}} = q\sqrt{\pi b}, \quad K_{\mathrm{III}} = q\sqrt{\pi b}, \tag{6.92}$$

where q is the shearing force intensity.

Similar SIF values for filled cracks are available from the solutions of respective problems [6]:

$$K_{\mathrm{II}} = \frac{q\sqrt{\pi b}(1 - \omega_1)}{1 + \beta\omega_1(1 - v)}, \quad K_{\mathrm{III}} = \frac{q\sqrt{\pi b}(1 - \omega_1)}{1 + \beta\omega_1}. \tag{6.93}$$

Figure 6.28a plots the dependence of K_{II} on the eccentricity a/b predicting by Eq. (6.75) for the case when shear direction coincides with longer ellipse semi-axis (a), i.e., in Eq. (6.74) $\alpha = 0$.

SIF decreases with growing parameter a/b. Such a shape of the K_{II} curve is observable only in the spatial problem definition.

Similar dependence of K_{III} on a/b under shear at $\varphi = \pi/2$ (at $\varphi = 0$, $K_{\mathrm{III}} = 0$) is shown in Fig. 6.28b). This curve, just as in the case of cleavage cracks, approaches SIF value corresponding to the plane problem. At $a/b \geq 10$, results for three-dimensional and two-dimensional problems differ negligibly.

In the case when shear direction coincides with shorter ellipse semi-axis b ($\alpha = \pi/2$ in Eqs. (6.74)), the dependence of SIF K_{II} and K_{III} on a/b is determinable from Eqs. (6.74) as before. Results are similar too, with the only exception being that the asymptotic conversion from the spatial problem to the plane one at $a/b \to \infty$ will be present for K_{II}, as it was for K_{III} in the previous case.

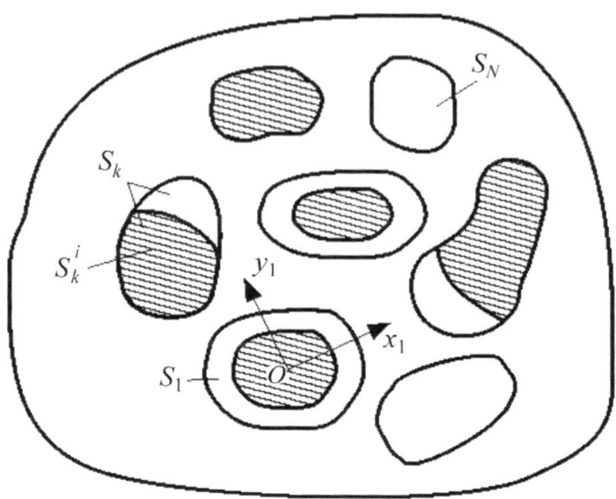

Fig. 6.29 Crack system in a body (cross-section with plane $z = 0$)

6.8   Injection Effect on the Strength of a Body Containing a System of Coplanar Cracks

Real materials contain, as a rule, systems of cracks with different sizes and orientation rather than single cracks. The technological processes of injection into defects must allow for interaction of these cracks. Let us consider a system of coplanar cracks oriented along a single plane of a deformable body and derive equations describing the interaction.

Problem Definition and Integral Solutions

Let an infinite body contain N plane cracks (S_k) restricted by smooth surfaces L_k $(k = 1,\ldots,N)$ (Fig. 6.29).

Assume that the injection has resulted in the filling (completely or partially) of N_1 cracks while other N_2 cracks remain empty.

Establish rectangular coordinate systems (x_k, y_k, z_k) with origins in areas S_k. Suppose that all cracks are located in one plane. Apply external tensile forces to the body perpendicular to the crack location plane such as to create normal uniform (containing no cracks) material stresses σ^0_{zzk} in the areas S_k $(k = 1,\ldots, N)$. Let us define the filler's response to the tensile load as similar to the one described by Eqs. (6.19). Based on the superposition principle valid in the linear elastic theory, reduce the problem of a body with N cracks to the following boundary problem for semi-space $z_k \geq 0$:

$$\sigma_{zzk}\Big|_{\bar{S}_k^i} = \frac{u_{zk} + u_{zk}^0}{h_k} E_k - \sigma_{zzk}^0 - \sum_{n=1(\pm k)}^{N} \sigma_{zzn},$$

$$\sigma_{zzk}\Big|_{\bar{S}_k^i} = -\sigma_{zzk}^0 - \sum_{n=1(\pm k)}^{N} \sigma_{zzn}, \qquad (6.94)$$

$$u_{zk}\Big|_{\bar{S}_k^i} = 0, \ \sigma_{xzk}\Big|_{S_k + \bar{S}_k} = \sigma_{yzk}\Big|_{S_k + \bar{S}_k} = 0.$$

Here S_k^i is a part of area S_k filled by the injection material; $\bar{S}_k^i = S_k - S_k^i$ is the empty part of the k-th crack; $\bar{S}_k$ is the area complementing the crack S_k till the plane $z_k = 0$; σ_{zzn} are stresses in the body with crack S_k.

An assumed absence of shear stresses in the plane $z_k = 0$ dictates that two of the three Papkovich-Neyber harmonic functions are exactly zero. The only non-zero function Φ allows for the expression of stress and displacement as follows:

$$\sigma_{zz} = -\frac{\partial^2 \Phi}{\partial z^2}, \qquad u_z = -\frac{1-\nu}{\mu}\frac{\partial \Phi}{\partial z} \qquad (6.95)$$

Selection of the harmonic function Φ in the form of the integral Fourier presentation (6.26) based on boundary conditions (6.94) produces a system of coupled integral equations for determining function $A_k(\xi,\eta)$:

$$\nabla^2 \int\int_{-\infty}^{\infty} \frac{\exp[i(x_k\xi + y_k\eta)]}{\sqrt{\xi^2 + \eta^2}} A_k(\xi,\eta)d\xi d\eta + \sum_{n=1(\pm k)}^{N} \int\int_{-\infty}^{\infty} A_n(\xi,\eta)\exp[i(x_k\xi + y_k\eta)]$$

$$\times\sqrt{\xi^2 + \eta^2}d\xi d\eta == \begin{cases} -\sigma_{zzk}^0 + \dfrac{u_{zk} + u_{zk}^0}{h_k} E_k, (x_k, y_k) \in S_k^i, \\[2mm] -\sigma_{zzk}^0, (x_k, y_k) \in \bar{S}_k^i, \end{cases}$$

$$\int\int_{-\infty}^{\infty} A_n(\xi,\eta)\exp[i(x_k\xi + y_k\eta)]d\xi d\eta = 0, (x_k, y_k) \in \bar{S}_k.$$

$$(6.96)$$

The latter of Eqs. (6.96) presents the condition connecting the edge displacements of each crack, which enables us to write it for the plane $z = 0$ in the form:

$$\frac{1-\nu}{\mu}\int\int_{-\infty}^{\infty} A_n(\xi,\eta)\exp[i(x_k\xi + y_k\eta)]d\xi d\eta = \begin{cases} u_{zk}, (x_k, y_k) \in S_k \\ 0, (x_k, y_k) \in \bar{S}_k \end{cases} \qquad (6.97)$$

Using the reversible Fourier transformation, we obtain from Eq. (6.94):

$$A_k(\xi,\eta) = \frac{\mu}{4\pi^2(1-\nu)}\int\int_{S_k} u_{zk}(x_k, y_k)\exp[-i(x_k\xi + y_k\eta)]dx_k dy_k. \qquad (6.98)$$

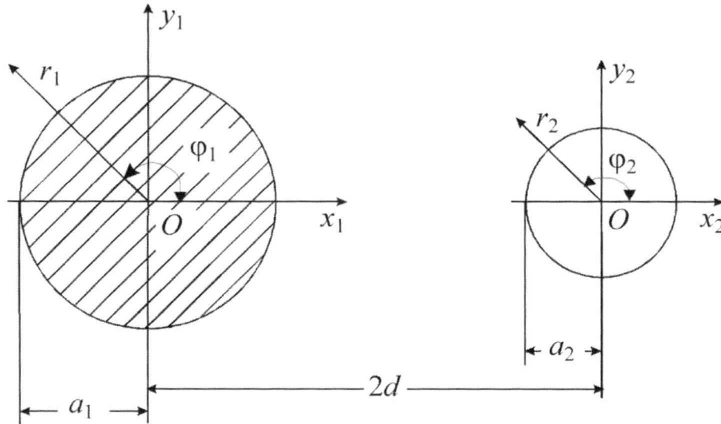

Fig. 6.30 Cross-section of a body containing two cracks (r_1 and r_2). One of the cracks (r_1) is filled with injection material, and the other (r_2) is empty

Now, substitute Eq. (6.98) into the first of Eqs. (6.96). Performing necessary transformations to determine the unknown crack edge displacements $u_{zk}\,(x_k,\,y_k)$, we come to the system of integral equations

$$\nabla^2 \int\limits_{-\infty}^{\infty} \int \frac{u_{zk}(x,y)dxdy}{\sqrt{(x_k-x)^2+(y_k-y)^2}} + \sum_{n=1(\pm k)}^{N} \int\limits_{S_n} \int \frac{u_{zn}(x,y)dxdy}{\sqrt{[(x_k-x)^2+(y_k-y)^2]^3}}$$

$$= \frac{4\pi(1-v^2)}{E} \left\{ \begin{array}{l} -\sigma_{zzk}^0 + \dfrac{u_{zk}+u_{zk}^0}{h_k} E_k, (x_k,y_k) \in S_k^i, \\[2ex] -\sigma_{zzk}^0, (x_k,y_k) \in \bar{S}_k^i, \end{array} \right.$$

$$k = 1,\ldots,N.$$
$$(6.99)$$

Solutions of the derived system (6.99) yield displacements u_{zk} in the crack area S_k. Thereafter, SIF values $K_1^{(k)}$ can be found using Eqs. (6.32).

Two Circular Cracks in a Deformable Body, one Filled by Injection Material

Let there exist two circular cracks with radii a_1 and a_2 in an infinite body, one (e.g., a_1) being filled with injection material (Fig. 6.30). A tensile load with intensity p acts on the body perpendicularly to the crack plane from the infinity. Suppose that crack surfaces V_1 and V_2 are ellipsoids of revolution with semi-axes a_1, a_2 and b_1, b_2 ($a_1 \gg b_1; a_2 \gg b_2$).

Determine the strength of the body containing such defects.

The present definition enables us to reduce the system of integral equations (6.99) to the system of two equations:

$$
\begin{cases}
\nabla^2 \iint\limits_{S_1} \dfrac{u_{z1}(x,y)\,dx\,dy}{\sqrt{(x_1-x)^2+(y_1-y)^2}} + \iint\limits_{S_2} \dfrac{u_{z2}(x,y)\,dx\,dy}{\sqrt{[(x_2-x)^2+(y_2-y)^2]^3}} \\[4mm]
\qquad = \dfrac{4\pi(1-v^2)}{E}\left(\dfrac{u_{z1}+u_{z1}^0}{h_1}E_1 - p\right), \quad (x_1,y_1)\in S_1; \\[6mm]
\nabla^2 \iint\limits_{S_2} \dfrac{u_{z2}(x,y)\,dx\,dy}{\sqrt{(x_2-x)^2+(y_2-y)^2}} + \iint\limits_{S_1} \dfrac{u_{z1}(x,y)\,dx\,dy}{\sqrt{[(x_2-x)^2+(y_2-y)^2]^3}} \\[4mm]
\qquad = -\dfrac{4\pi(1-v^2)p}{E}, \quad (x_2,y_2)\in S_2.
\end{cases}
\tag{6.100}
$$

We shall seek the continuous solution of system (6.100) in the form:

$$
\begin{aligned}
u_{zk}(x_k,y_k) = \sqrt{a_k^2 - x_k^2 - y_k^2}\,(A_{0k} + A_{1k}x_k + A_{2k}y_k \\
+ A_{3k}x_k^2 + A_{4k}x_k y_k + A_{5k}y_k^2), k = 1,2.
\end{aligned}
\tag{6.101}
$$

Here, A_{jk} ($j=0\ldots5$) are unknown coefficients.

Substitution of the expression (6.101) into Eq. (6.100), expansion of the regular kernel components in a series using $\lambda_k = a_k/d < 1$, where $2d$ is the distance between crack centers, as a small parameter, and making some necessary transformations results in the following expressions for coefficients A_{jk}:

$$
\begin{cases}
A_{01} = \dfrac{4p(1-\omega_1)(1-v^2)}{E[\pi+4(1-v^2)\omega_1\beta_1]} + \dfrac{2p(1-v^2)}{3E[\pi+4(1-v^2)\omega_1\beta_1]\pi} \\[3mm]
\qquad \times \left(\dfrac{\lambda_2^3}{2} + \dfrac{\lambda_1^2\lambda_2^3}{16} + \dfrac{9\lambda_2^5}{80} - \dfrac{\lambda_1^2\lambda_2^3(1-v^2)\omega_1\beta_1}{9\pi+16(1-v^2)\omega_1\beta_1}\right); \\[4mm]
A_{02} = \dfrac{4p(1-v^2)}{E\pi} + \dfrac{2p(1-v^2)(1-\omega_1)}{3E\pi[\pi+4(1-v^2)\omega_1\beta_1]}\left(\dfrac{\lambda_1^3}{2} + \dfrac{\lambda_2^2\lambda_1^3}{16} + \dfrac{9\lambda_1^5}{80}\right); \\[4mm]
A_{11} = \dfrac{p(1-v^2)\lambda_2^3}{E\pi l[3\pi+8(1-v^2)\omega_1\beta_1]}; \quad A_{12} = \dfrac{p(1-\omega_1)(1-v^2)\lambda_1^3}{3E\pi l[\pi+4(1-v^2)\omega_1\beta_1]}; \\[4mm]
A_{21} = A_{22} = A_{41} = A_{42} = 0; \\[4mm]
A_{31} = \dfrac{15p(1-v^2)\lambda_2^3[135\pi+256(1-v^2)\omega_1\beta_1]}{4E\pi l^2\left\{[45\pi+88(1-v^2)\omega_1\beta_1]^2 - 64(1-v^2)\omega_1^2\beta_1^2\right\}}; \\[4mm]
A_{32} = \dfrac{p(1-\omega_1)(1-v^2)\lambda_1^3}{4E\pi[\pi+4(1-v^2)\omega_1\beta_1]l^2}; \\[4mm]
A_{51} = -\dfrac{15p(1-v^2)\lambda_2^3[45\pi+64(1-v^2)\omega_1\beta_1]}{4E\pi l^2\left\{[45\pi+88(1-v^2)\omega_1\beta_1]^2 - 64(1-v^2)\omega_1^2\beta_1^2\right\}}; \\[4mm]
A_{52} = -\dfrac{p(1-\omega_1)(1-v^2)\lambda_1^3}{12E\pi l^2[\pi+4(1-v^2)\omega_1\beta_1]}; \quad \beta_1 = \dfrac{a_1}{c_1}.
\end{cases}
\tag{6.102}
$$

The coefficients (6.102) and Eq. (6.101) lead to the following formulas for SIF near each crack:

$$K_{\mathrm{I}}^{(1)} = 2p\sqrt{\pi a_1}\left\{\frac{1-\omega_1}{\pi+4\left(1-\nu^2\right)\omega_1\beta_1}+\frac{\lambda_2^3}{\pi}\left[\frac{1}{\pi+4\left(1-\nu^2\right)\omega_1\beta_1}\right.\right.$$

$$\times\left(\frac{1}{12}\left(1+\frac{\lambda_1^2}{8}+\frac{9\lambda_2^2}{40}\right)-\frac{\lambda_1^2\left(1-\nu^2\right)\omega_1\beta_1}{6\left[9\pi+16\left(1-\nu^2\right)\omega_1\beta_1\right]}\right)$$

$$+\frac{\lambda_1\cos\varphi_1}{4[3\pi+8(1-\nu^2)\omega_1\beta_1]}$$

$$+\frac{15\left[135\pi+256\left(1-\nu^2\right)\omega_1\beta_1\right]\lambda_1^2\cos^2\varphi_1}{16\left\{\left[45\pi+88\left(1-\nu^2\right)\omega_1\beta_1\right]^2-64(1-\nu^2)^2\omega_1^2\beta_1^2\right\}}$$

$$\left.\left.-\frac{15\left[45\pi+64\left(1-\nu^2\right)\omega_1\beta_1\right]\lambda_1^2\sin^2\varphi_1}{16\left\{\left[45\pi+88\left(1-\nu^2\right)\omega_1\beta_1\right]^2-64(1-\nu^2)^2\omega_1^2\beta_1^2\right\}}\right]\right\};$$

$$K_{\mathrm{I}}^{(2)} = 2p\sqrt{\pi a_2}\left\{\frac{1}{\pi}+\frac{(1-\omega_1)\lambda_1^3}{\pi+4\left(1-\nu^2\right)\omega_1\beta_1}\left[\frac{1}{\pi}\times\right.\right.$$

$$\left.\left.\times\frac{1}{12}\left(1+\frac{\lambda_1^2}{8}+\frac{9\lambda_2^2}{40}\right)-\frac{\lambda_2\cos\varphi_2}{12\pi}+\frac{\lambda_2^2\cos^2\varphi_2}{16}-\frac{\lambda_2^2\sin^2\varphi_2}{\pi}\right]\right\}.$$

$$(6.103)$$

Here φ_k is the polar angle in the polar coordinate system r_k, φ_k with origin in the center of area S_k.

Maximum values of the intensity factors and, hence, the initial propagation place of the most dangerous crack depend on crack sizes, filler's hardness $\omega_1 = E_1/E$, and crack opening parameter β_1. Depending on the balance of these parameters, fracture can begin from either an empty or filled crack. One can analyze the possible scenarios using formulae (6.103). The strength of a body with two cracks (Fig. 6.30), one filled by injection material and another empty, is subject to determination using formulae (6.103) and criterion (6.5). The result of the problem under consideration consists of the following expression:

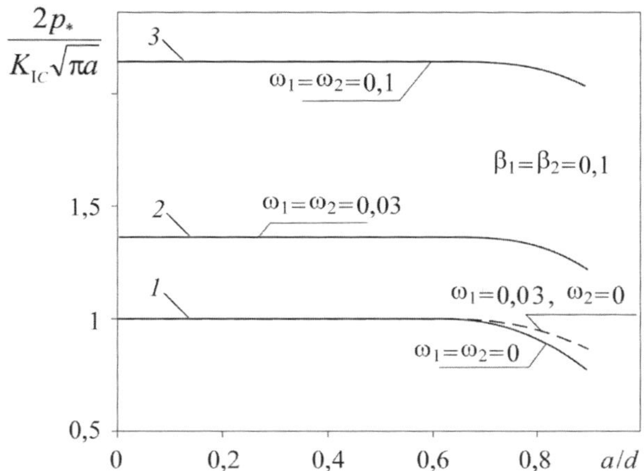

Fig. 6.31 Dependence of ultimate load on parameter d (distance between centers of two equal circular and coplanar cracks) at various conditions of filling

$$
p_* = \frac{K_{IC}}{2\sqrt{\pi}}\max\left\{
\begin{aligned}
&\sqrt{a_1}\left\{\frac{1-\omega_1}{\pi+4(1-\nu^2)\omega_1\beta_1} + \frac{\lambda_2^3}{\pi}\left[\frac{1}{\pi+4(1-\nu^2)\omega_1\beta_1}\right.\right.\\
&\left(\frac{1}{12}\left(1+\frac{\lambda_1^2}{8}+\frac{9\lambda_2^2}{40}\right)\right.\\
&\left.-\frac{\lambda_1^2(1-\nu^2)03.\omega_1\beta_1}{6[9\pi+16(1-\nu^2)\omega_1\beta_1]}\right)+\frac{\lambda_1}{4[3\pi+8(1-\nu^2)\omega_1\beta_1]}\\
&\left.\left.+\frac{15[135\pi+256(1-\nu^2)\omega_1\beta_1]\lambda_1^2}{16\left\{[45\pi+88(1-\nu^2)\omega_1\beta_1]^2-64(1-\nu^2)^2\omega_1^2\beta_1^2\right\}}\right]\right\};\\
&\sqrt{a_2}\left\{\frac{1}{\pi}+\frac{(1-\omega_1)\lambda_1^3}{\pi+4(1-\nu^2)\omega_1\beta_1}\right.\\
&\left.\left[\frac{1}{12\pi}\left(1+\frac{\lambda_1^2}{8}+\frac{9\lambda_2^2}{40}\right)+\frac{\lambda_2}{12\pi}+\frac{\lambda_2^2}{16}\right]\right\}^{-1}.
\end{aligned}
\right.
$$

$$\tag{6.104}$$

Figure 6.31 shows some extreme cases of limit equilibrium in a body with two circular coplanar cracks under tension. In particular, it shows p_* for two empty cracks with identical radii (curve *1*) as a function of crack spacing (d) and similar dependences for filling of one crack (dashed line) or both cracks with injection materials of different hardness (curves *2, 3*).

6.9   Limit Equilibrium in a Plate with Two Filled Cracks [28]

Let us consider the plate with two cracks of identical length ($2l$) under tension by load intensity p perpendicularly to cracks occurring in the plane (Fig. 6.32). Designate the distance between crack centers as $2d$. We are seeking the limit equilibrium state in this plate for either empty cracks or those filled with injection material.

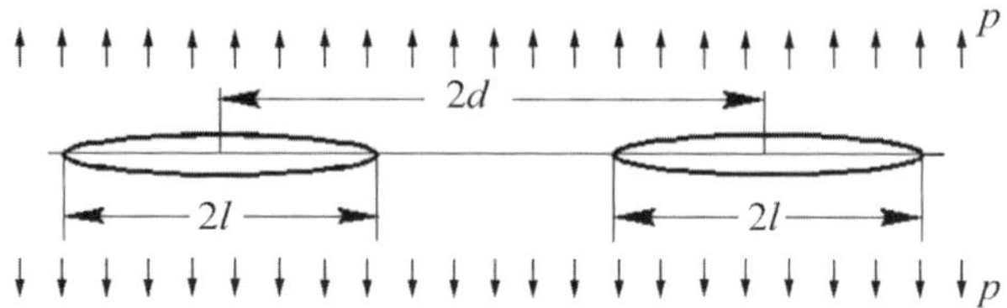

Fig. 6.32 Layout of two cracks in the plate

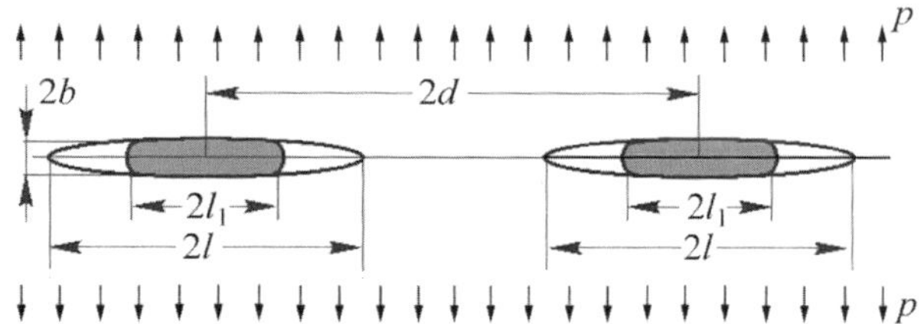

Fig. 6.33 Filling pattern for two collinear cracks

SIF for free (empty) cracks is available using formula [22]:

$$K_{\mathrm{I}}^{\pm} = \pm \frac{p\sqrt{\pi l}}{\lambda} \frac{1 \pm \lambda - E(\lambda)/K(\lambda)}{\sqrt{1 \pm \lambda}}, \quad \lambda \equiv \frac{l}{d}. \tag{6.105}$$

Signs $(+)$ or $(-)$ correspond to the inner or outer crack tips, respectively.

The inner crack tips reach the limit equilibrium state first. We can estimate the ultimate load value p_* for this case from Eqs. (6.15) and (6.105):

$$p_* = \frac{K_{\mathrm{IC}}}{\sqrt{\pi l}} \frac{\lambda \sqrt{1 + \lambda}}{1 + \lambda - E(\lambda)/K(\lambda)}. \tag{6.106}$$

Example 1 Assume the two above-illustrated cracks to be filled with injection material in the manner presented below in Fig. 6.33.

One can derive the describing integral equation from the system (6.99) by integrating the equations with respect to the variable η from $-\infty$ to ∞ and putting the following values for partial integrals:

$$\left(\frac{\partial^2}{\partial x^2} + \frac{\partial^2}{\partial y^2} \right) \int_{-\infty}^{\infty} \frac{d\eta}{\sqrt{(x - \xi)^2 + (y - \eta)^2}} = \frac{2}{(x - \xi)^2};$$

$$\frac{\partial^2}{\partial \eta^2} \int_{-\infty}^{\infty} \frac{d\eta}{\sqrt{(x - \xi)^2 + (y - \eta)^2}} = 0.$$

Such a procedure yields the following singular integro-differential equation in respect to unknown crack edge displacements $u_y(x)$:

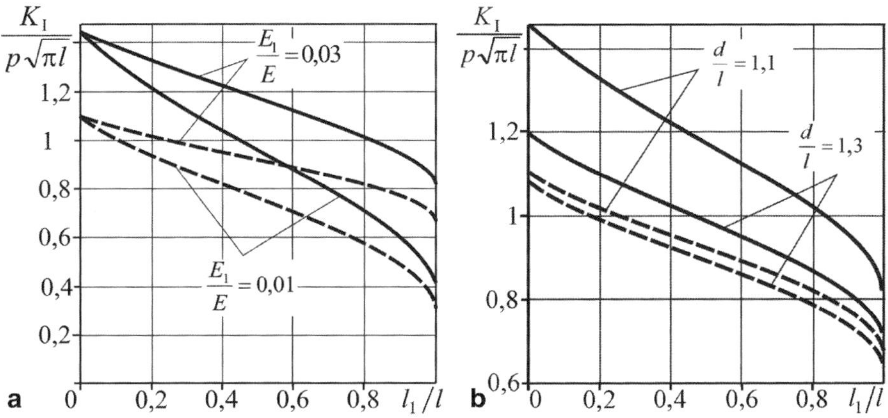

Fig. 6.34 SIF dependence on filling extent (l_1/l) at various values of filler hardness E_1/E (**a**) and crack spacing l/b (**b**)

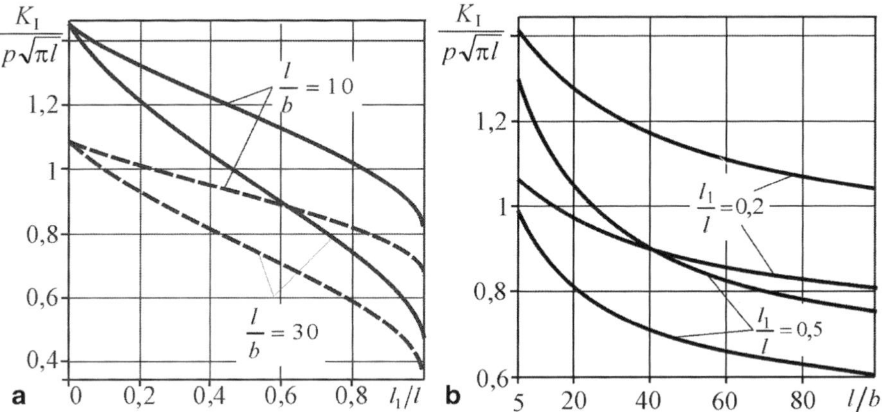

Fig. 6.35 SIF dependence on filling extent (**a**) and crack opening (**b**) ($E_1/E = 0.03$; $d/l = 1.1$)

$$\int_{-l}^{l} \frac{(t+d)u_y'(t)dt}{(t-x)(t+x+2d)} = \frac{\pi(1-v^2)}{E}$$

$$\left(\frac{u_y(x)E_1}{h(x)} H\left(l_1 - |x|\right) + p\left(\frac{E_1}{E} H(l_1 - |x|) - 1 \right) \right). \tag{6.107}$$

Let us solve this equation numerically using Chebyshev–Gauss quadratures. Accept a defect shape elliptical with semi-axes l and b ($l \gg b$).

Figures 6.34, 6.35, 6.36 and 6.37 present results of the numerical analysis. Here, dashed lines denote outer crack tips and solid lines inner crack tips.

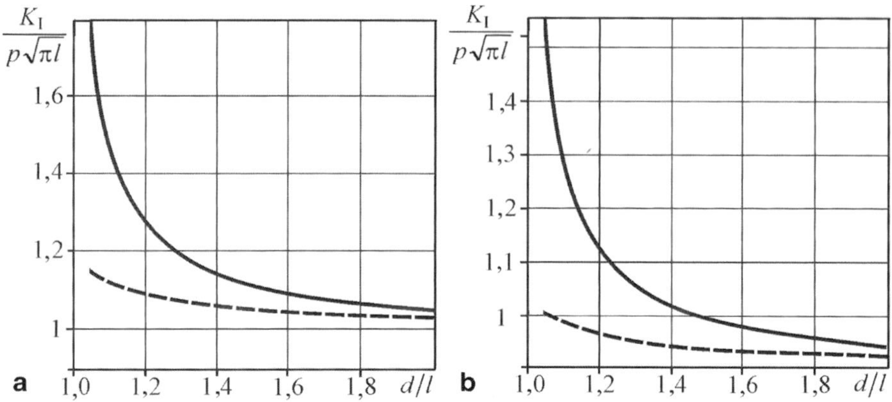

Fig. 6.36 SIF dependence on crack spacing (d/l) for empty cracks (**a**) or cracks filled with injection material (**b**). Here $l/b = 10$; $l_1/l = 0.3$; $E_1/E = 0.03$

Fig. 6.37 SIF dependence on filler hardness E_1/E at various values of filling extent (l_1/l) ($l/b = 10$; $d/l = 1.1$)

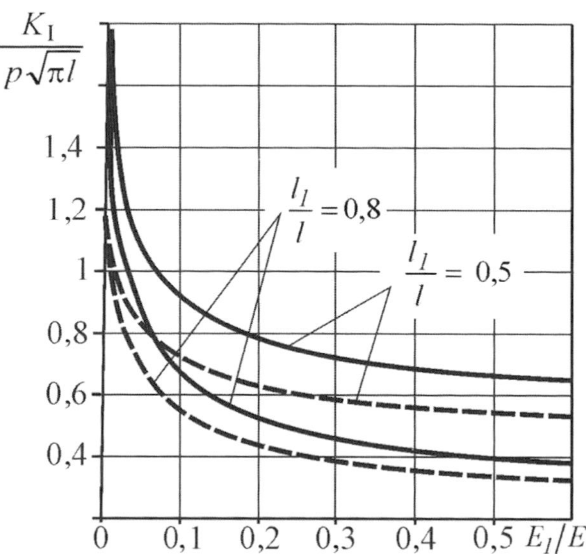

Example 2 Now, assume two cracks filled in the manner presented in Fig. 6.38.

The singular integral equation for this problem follows from the system (6.49) in the form

$$\int_{-l}^{l} \frac{(t+d)u'_y(t)dt}{(t-x)(t+x+2d)} = \frac{\pi(1-v^2)}{E}$$

$$\left(\frac{u_y(x)E_1}{h(x)} H\left(|x| - l_1\right) + p\left(\frac{E_1}{E} H(|x| - l_1) - 1\right) \right), \tag{6.108}$$

$$|x| \leq l.$$

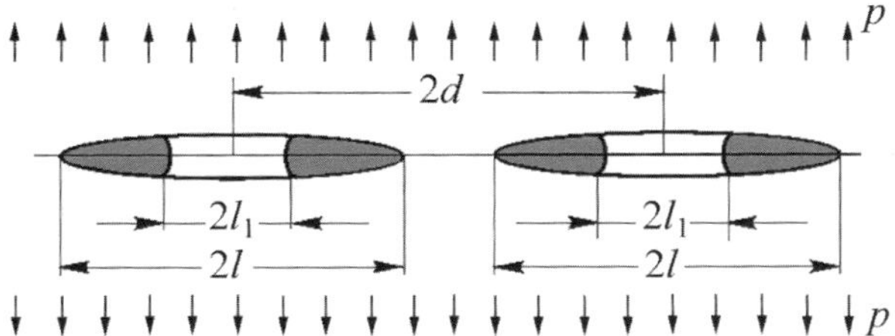

Fig. 6.38 Two cracks filled with injection material in the tips

Figures 6.39 and 6.40 present the numerical solutions of integral equation (6.108) at various values of parameters. Here, dashed lines again denote outer crack tips and solid lines inner crack tips.

Example 3 Let us consider the case of a plate with two identical cracks when the injection material fills the crack beginning from one side (Fig. 6.41).

The system of equations (6.99) in the given case leads to the following singular integro-differential equation in respect to unknown crack edge displacements u_y:

$$\int_0^{2l} \frac{(t + d - l)u'_y(t)dt}{(t - x)(t + x + 2d - 2l)} = \frac{\pi(1 - v^2)}{E}$$
$$\left(\frac{u_y(x)E_1}{h(x)} H(l_1 - x) + p\left(\frac{E_1}{E} H(l_1 - x) - 1 \right) \right),$$
$$0 \le x \le 2l. \tag{6.109}$$

Figures 6.42 and 6.43 present numerically calculated SIF values for inner (dashed lines) and outer crack tips (solid lines).

Example 4 Let us consider the case when the injection material fills the crack beginning from the inner sides (Fig. 6.44). The singular integro-differential equation in respect to unknown crack edge displacements u_y takes the appearance

$$\int_0^{2l} \frac{(t + d - l)u_y(t)dt}{(t - x)(t + x + 2d - 2l)} = \frac{\pi(1 - v^2)}{E}\left(\frac{u_y(x)E_1}{h(x)} H(x - l_1) + p \right.$$
$$\left. \left(\frac{E_1}{E} H(x - l_1) - 1 \right), \right.$$
$$0 \le x \le 2l. \tag{6.110}$$

Figures 6.45 and 6.46 present numerically calculated SIF values from Eq. (6.110) at various values of parameters.

Using plotted dependences of SIF on crack size, crack opening, crack spacing, filler hardness, relative size of injection-filled zone, and position of such zones in the cracks, one can quantitatively estimate the decrease in SIF due to injection and ensure selection of the best injection parameters providing required residual strength of the cracked body.

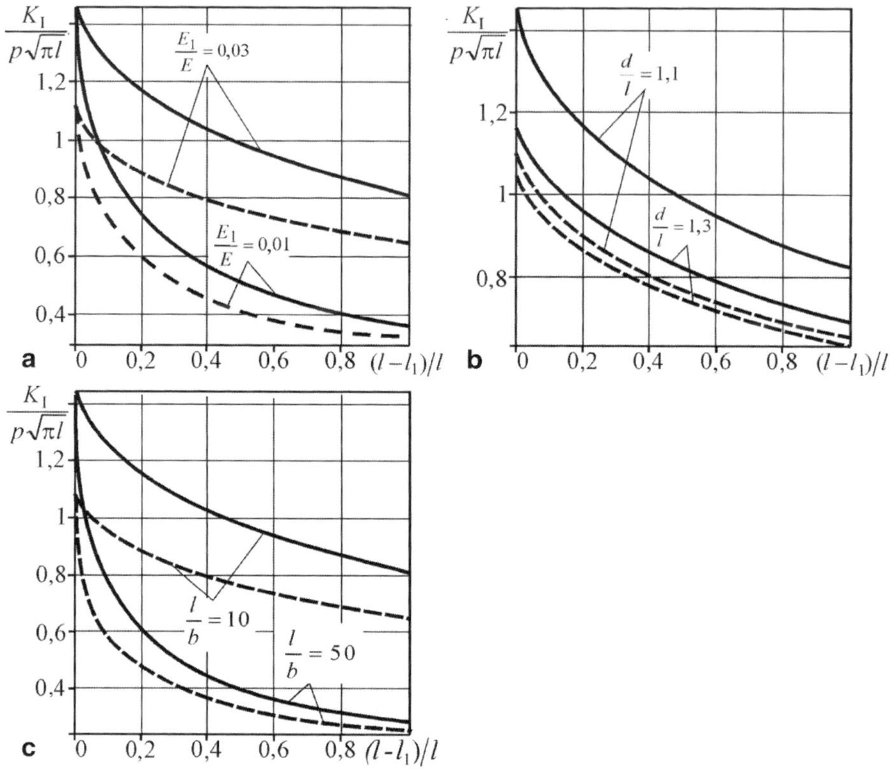

Fig. 6.39 SIF dependence on degree of crack filling $(l - l_1)/l$ at various values of filler hardness E_1/E (**a**), crack spacing d/l (**b**), and crack opening l/b (**c**)

Fig. 6.40 SIF dependence on crack spacing d/l at various values of degree of crack filling $(l - _1)/l$

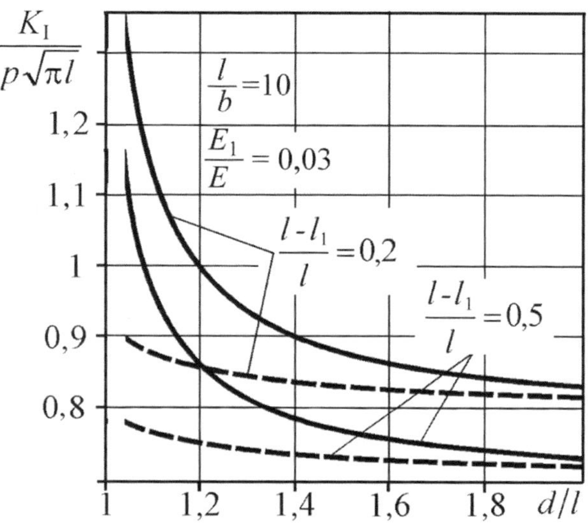

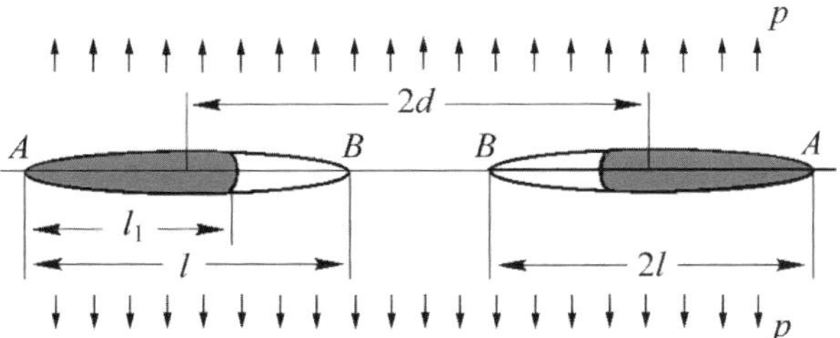

Fig. 6.41 Two cracks filled with injection material from the outer sides

6.10   Tension of a Plate with a Surface Crack [29]

Let a straight boundary crack with length l be located in an elastic semi-plane so that it forms the angle θ with the boundary (Fig. 6.47). The applied load acts as shown in the figure. The solution of this problem for the empty crack including SIF and ultimate loads p already exists in the literature [22].

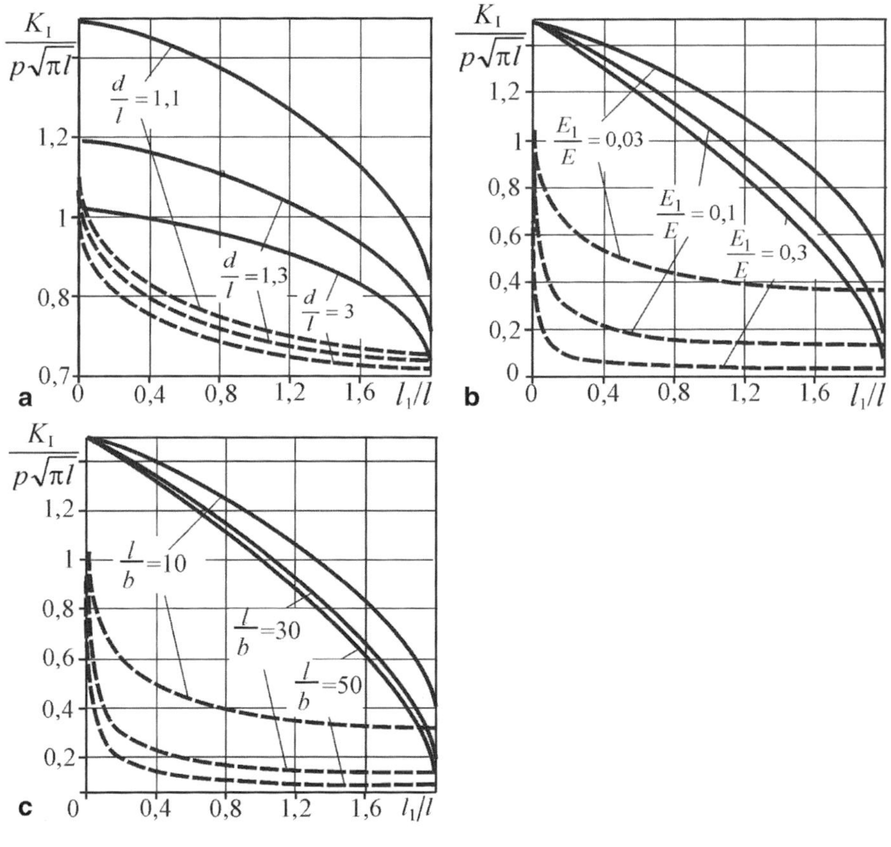

Fig. 6.42 SIF values for outer crack tips as functions of filling extent (l_1/l) at various values of crack spacing d/l (**a**), filler hardness E_1/E (**b**), and crack opening l/b (**c**)

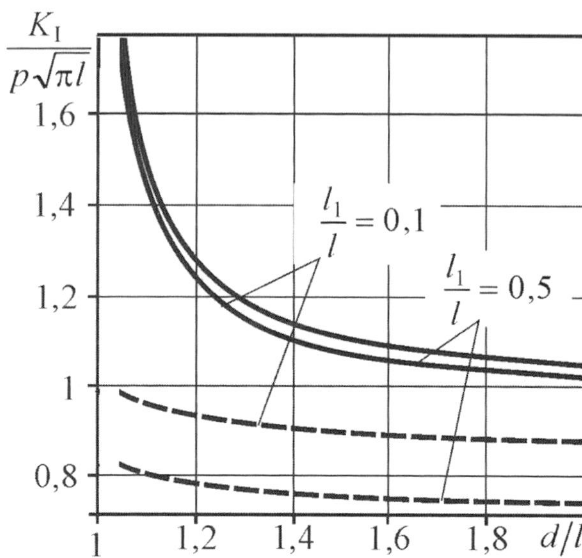

Fig. 6.43 SIF values for outer crack tips as functions of crack spacing d/l at various values of filling extent (l_1/l)

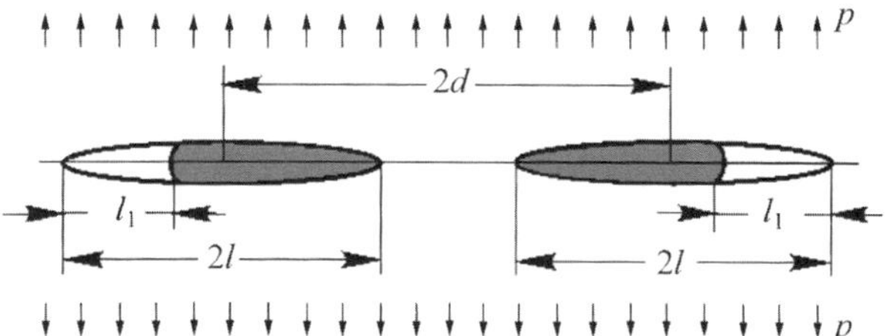

Fig. 6.44 Two cracks filled with injection material from the inner sides

Let us now consider the crack partially filled with injection material beginning from its tip. Based on the model of the Winkler foundation [6], we can reduce this problem to the singular integro-differential equation in respect to unknown crack edge displacements $u_y\,(x)$ [29]:

$$\int_{-1}^{1} \left\{ u'_y(\xi) \left[\frac{1}{\xi - \eta} + \frac{1}{2} \left(\frac{1}{\eta - \xi e^{2i\theta}} + \frac{\eta^2 + \xi \eta (1 - 4e^{-2i\theta} + e^{-4i\theta})}{(\eta - \xi e^{2i\theta})^3} \right) \right] \right.$$

$$\left. + \bar{u}'_y(\xi) \frac{\xi}{2} \left[\frac{1 - e^{-2i\theta}}{(\eta - \xi e^{2i\theta})^2} + \frac{e^{-2i\theta}(1 - e^{-2i\theta})}{(\eta - \xi e^{2i\theta})^2} \right] \right\} d\xi = \frac{2\pi(1 - \nu^2)}{E} p(\eta), \quad |\eta| < 1.$$

$$(6.111)$$

where $p(\eta) = -pl(1 - \exp(-2i\theta))/2$ or $p(\eta) = -pl(1 - E_1/E - u_y E_1/hp)$ $(1 - \exp(-2i\theta))/2$ in the empty or filled parts of the crack, respectively.

Figure 6.48 shows the results of a numerical solution at $\theta = \pi/2$. We can see an abrupt SIF drop at the filling of even an insignificant part of the crack near its tip. An empty crack with $K_1 = 1.12p\sqrt{\pi l}$ is designated by an asterisk in this plot.

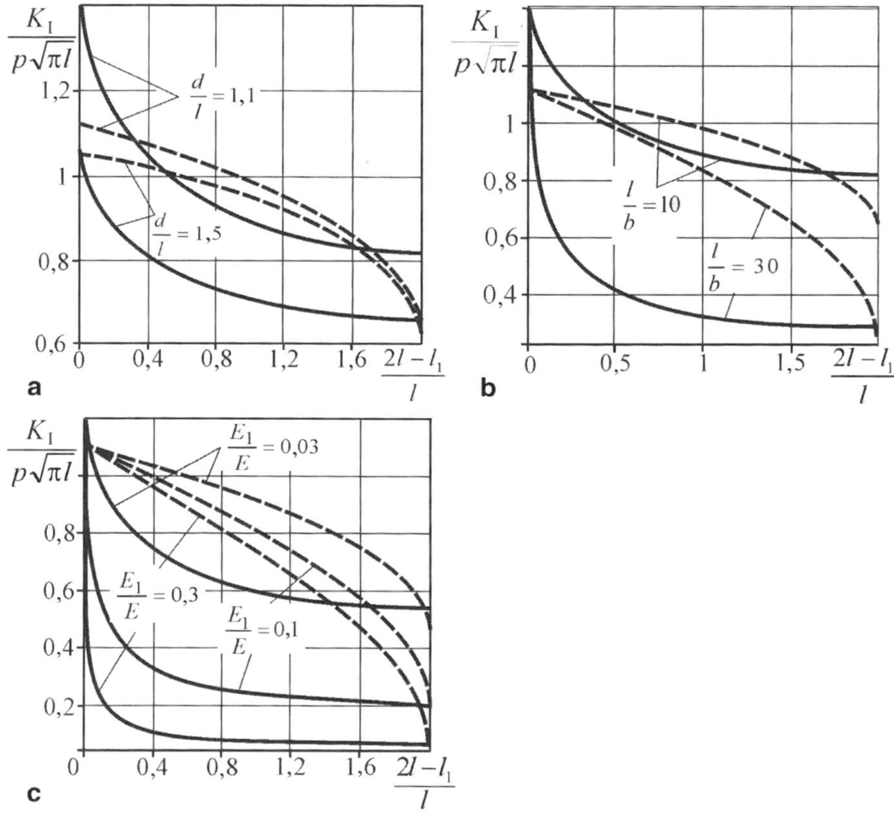

Fig. 6.45 SIF values as functions of filling extent (l_1/l) at various values of crack spacing d/l (**a**), crack opening l/b (**b**), and filler hardness E_1/E (**c**)

An appropriate selection of filler with optimal hardness (E_1/E) provides an essential decrease in SIF near the boundary crack (Fig. 6.48a), and, consequently, essential strength restoration in the body damaged by the crack. The above presented data (Fig. 6.48b) indicate that it is preferable to treat cracks with the highest possible parameter l/b, i.e., the smallest initial opening in order to provide the best healing.

6.11   Renewal of Load-Carrying Capability for a Disk with a Central Crack Under Compressive Loading Along the Crack [30]

Engineering practice often encounters the necessity of restoring the load-carrying capacity of structural elements made from brittle materials such as ceramics, concrete, rocks, hard alloys, etc., that work under compressive loading.

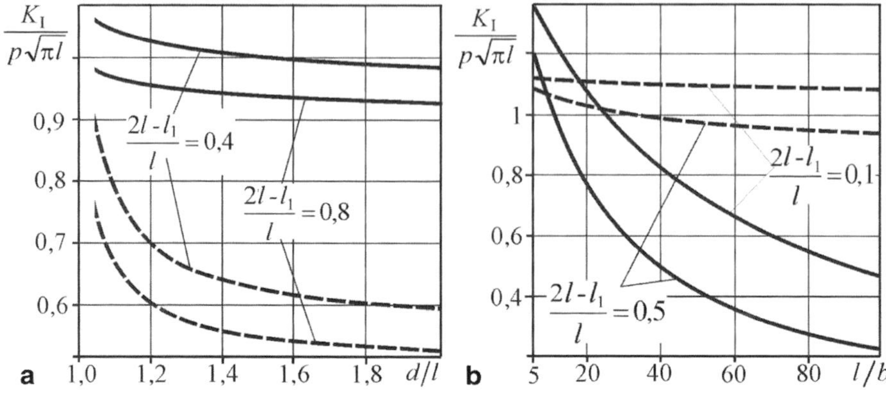

Fig. 6.46 Dependence of SIF on crack spacing d/l (**a**) and crack shape l/b (**b**) at various values of degree of crack filling $(2l - l_1)/l$

Fig. 6.47 Arbitrarily oriented
surface crack

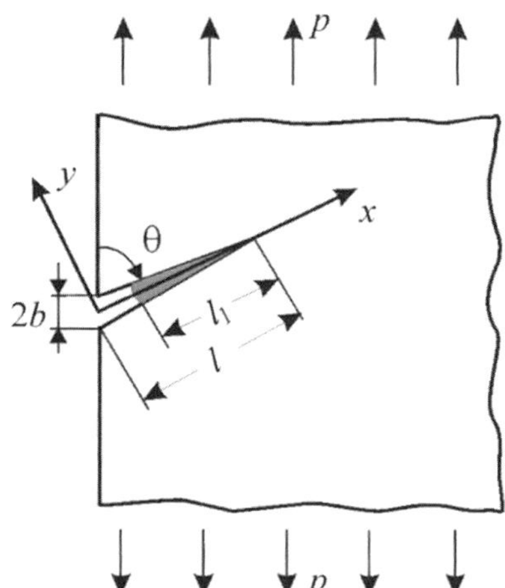

Let us consider a disk damaged by a central crack with length $2l$ (disk radius R, disk thickness t) loaded by force Q (Fig. 6.49) as an example of the above element. We shall assume the load Q uniformly distributed along the disk generatrix with intensity $p = Q/t$.

In such an arrangement, tensile stresses σ_{yy} arise near the crack tips with an intensity factor defined as [24]:

$$K_1 = p\sqrt{\frac{2\lambda}{\pi(1 - \lambda^2)}}\,f(\lambda),$$

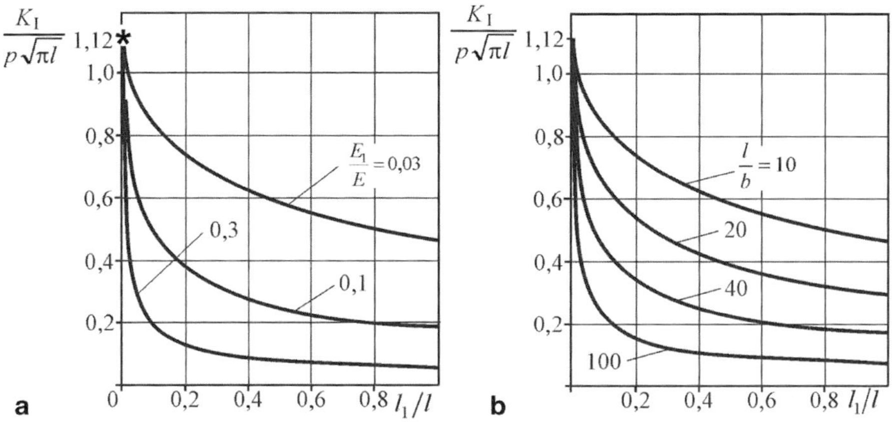

Fig. 6.48 SIF values as functions of extent of crack filling l_1/l at various values of filler hardness E_1/E (**a**) and crack opening l/b (**b**)

Fig. 6.49 Geometry and loading scheme for a disk with a crack

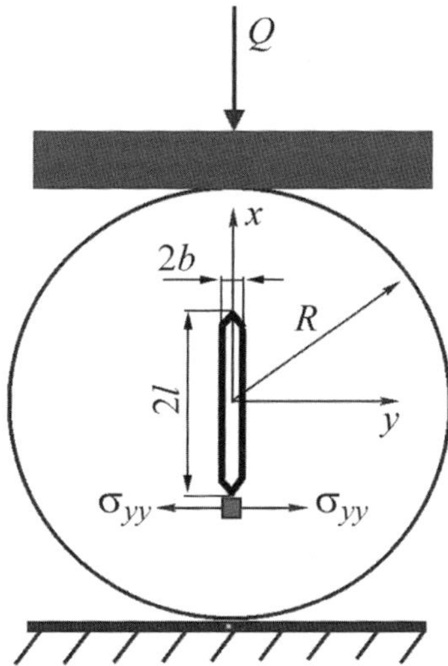

where

$$f(\lambda) = 1 + \lambda^2 - \frac{7}{8}\lambda^4 + \frac{1}{2}\lambda^6 - \frac{59}{128}\lambda^8 + \frac{1}{128}\lambda^{10} - 0.3447\lambda^{12};$$

$$\lambda \equiv \frac{l}{R}.$$

(6.112)

At $\lambda < 0.9$, the calculation error here is less than 0.2 %.

The ultimate load value ($p = p_*$) for such a cracked disk is determined by the criterion $K_1(p_*,l) = K_{IC}$. This gives

$$p_* = K_{IC} \left\{ \sqrt{\frac{2\lambda}{\pi(1 - \lambda^2)}} f(\lambda) \right\}^{-1}, \tag{6.113}$$

$$K_{IC} = p_* \sqrt{\frac{2\lambda}{\pi(1 - \lambda^2)}} f(\lambda). \tag{6.114}$$

Equation (6.113) allows for calculation of the ultimate (highest permissible) load for a disk with a crack (Fig. 6.49) based on known values K_{IC}, R, and l. At the same time, it is useful to determine K_{IC} from experimentally measured p_* values[1].

Let us now assume that there is a material inside the crack-like defect or acute slit in the disk connecting the edges of the crack or slit after hardening. In such a case, the renewed element (disk) is able to carry a load p higher than p_* estimated from Eq. (6.113). We have to establish the ultimate load $p = p_{**}$ for this case that accounts for the mechanical properties of the filler.

With this purpose, we reduce the boundary problem describing the loading configuration of the disk with the filled crack to the following singular integro-differential equation in respect to unknown crack edge displacements in the section $-l \leq x \leq l$ [30] (Fig. 6.49):

$$\frac{E}{2\pi(1 - v^2)} \int_{-1}^{1} \left(\frac{1}{\eta - \xi} + M(\eta, \xi) \right) u'_y(\eta) d\eta - \frac{u_y(\xi)}{h(\xi)} E_1 = \frac{p(\omega - 1)}{\pi R}, \tag{6.115}$$

where

$$M(\eta, \xi) = \frac{\lambda^2}{2(1 - \lambda^2 \eta \xi)^3} [4\xi - 5\eta + \lambda^2 \eta(\xi\eta + 3\eta^2 - 3\xi^2) + \lambda^4 \xi \eta^2(\xi\eta$$

$$+ \xi^2 - \eta^2)] + \frac{\lambda^2}{2(1 - \lambda^2 \xi \eta)^2} [2\xi - \eta + \lambda^2 \eta(\eta^2 - 2\eta\xi - \xi^2) + \lambda^4 \eta^3 \xi^2];$$

$\omega = E_1/E$; $\xi = x/l$; E and v are the Young modulus and Poisson ratio of the host material, respectively. It is intended here that $u^* = u + u_0$; $u_0 = ph/\pi RE$. This equation is solvable only numerically at arbitrary function $h(x)$.

However, we can find an asymptotic analytical solution, if we assume that the slot has an elliptical shape with semi-axes l and b ($l \gg b$). Let us seek such a solution in the form

$$u_y(\xi) = \sqrt{1 - \xi^2}(A_0 + A_1\xi^2 + A_2\xi^4), \tag{6.116}$$

---

[1] This formula is especially useful in engineering practice for measuring the fracture toughness of brittle materials such as concretes, ceramics, hard alloys, etc. [24].

which gives the possibility of having the solution to within quantities as small as λ^4. Here, A_0, A_1, and A_2 are unknown coefficients.

Substituting the expression (6.116) into Eq. (6.115) and equating coefficients at identical powers of the variable ξ, we get the constants A_0, A_1, and A_2 as follows:

$$\begin{cases} A_0 = \dfrac{2(1-v^2)p}{\pi RE}\dfrac{1-\omega}{\beta_1}\left[1 + \dfrac{3}{2}\dfrac{1}{\beta_1}\lambda^2 + \dfrac{3}{4}\left(\dfrac{1}{2+\beta_1} - \dfrac{2}{\lambda} + \dfrac{3}{\beta_1^2}\right)\lambda^4\right]; \\[2ex] A_1 = \dfrac{2(1-v^2)p}{\pi RE}\dfrac{1-\omega}{\beta_1}\dfrac{3}{2+\beta_1}\lambda^4; \\[2ex] A_2 = 0;\; \beta = \dfrac{l}{b};\; \beta_1 = 1 + 2\beta\omega(1-v^2). \end{cases}$$

$$(6.117)$$

Now, we can estimate a stress intensity factor K_1 that characterizes the stress state near the filled crack using the approximation:

$$K_1 = \left(\frac{-E\sqrt{\pi l}}{2(1-v^2)}\right) \cdot \lim_{\xi \to 1}\sqrt{1 - \xi^2}u_y'(\xi). \qquad (6.118)$$

From here, accounting for Eqs. (6.116) and (6.117), we come to the expression:

$$K_1 = \left(\frac{p}{R}\sqrt{\frac{\pi}{l}}\right)\frac{1-\omega}{\beta_1}[1 + B_0\lambda^2 + B_1\lambda^4], \qquad (6.119)$$

where

$$B_0 = \frac{3}{2}\frac{1}{\beta_1}, \quad B_1 = \frac{3}{4}\left(\frac{-3}{2+\beta_1} - \frac{2}{\beta_1} + \frac{3}{\beta_1^2}\right)$$

Thus, Eq. (6.119) permits for the calculation of the highest allowable load applicable to the renewed structural element without fracture.

In this view, one must suppose that the filler's strength and adhesion to the host material are high enough to ensure local fracture origin in the vicinity of the crack tip, and that the limit equilibrium state takes place at $K_1(p_*, l) = \tilde{K}_{IC}$, where $\tilde{K}_{IC}$ reflects resistance to the filled crack growth. Then, the ultimate load p_{**} results from Eq. (6.119) as:

$$p_{**} = \tilde{K}_{IC}R\sqrt{\frac{\pi}{l}}\left(\frac{1-\omega}{\beta_1}\right)^{-1}[1 + B_0\lambda^2 + B_1\lambda^4]^{-1}. \qquad (6.120)$$

The parameter $\tilde{K}_{IC}$, generally speaking, can differ from respective fracture toughness constant K_{IC} and is subject to experimental measurement. Such experiments are necessary elements in building mathematical strength models for materials and structures with defects similar to filled cracks. In the present work, we have carried out experiments to measure values $\tilde{K}_{IC}$ based on Eq. (6.119). We have established that, in the case of concrete as the host material and polyurethane as the filler, the parameter of filled crack growth resistance $\tilde{K}_{IC}$ coincides with the fracture toughness K_{IC}.

Fig. 6.50 Concrete specimen with the filled slit before (**a**) and after (**b**) testing

Table 6.1 Experimental test data

Specimen type	Ultimate load Q_*, kN
Without slit	36... 45
With slit	15.8... 17.5
With filled slit	28.4... 30.4

The ratio of parameters expressed by Eqs. (6.113) and (6.120) illustrates the strengthening effect in the cylindrical structural element damaged by the crack and restored by injection:

$$\frac{p_*}{p_{**}} = \left(\frac{1-\omega}{1+2\beta\omega(1-\nu^2)}\right)[1 + B_0\lambda^2 + B_1\lambda^4]\left[1 + \frac{3}{2}\lambda^2\right]^{-1}. \qquad (6.121)$$

We made a series of fifteen concrete cylindrical specimens of $t = 100$ mm in height and $D = 100$ mm in diameter. The concrete mix was prepared from the cement brand 500 of the Mykolayev cement plant and sand fraction 0.6... 1.0 mm in a proportion of 1:3. The specimens were jolt molded at a frequency of 50 Hz over five minutes.

Through slits were formed in ten specimens (Fig. 6.50a) using metal inserts with cross-sections close to the ellipse with semi-axes $l = 20$ mm, $b = 2$ mm, and a radius of curvature in the tips less than 0.1 mm. The formed specimens with and without slits were exposed to room conditions over 30 days. After the exposure, we filled the slits in five of the specimens with polyurethane. The Young moduli were 360 MPa for polyurethane and 12000 MPa for concrete. The Poisson ratio for concrete was $\nu = 0.2$. Table 6.1 below presents the test data in the configuration shown in Fig. 6.49. The measured parameters enabled the calculation of the ultimate tensile strength of concrete σ_B, fracture toughness of concrete K_{IC}, and parameter $\tilde{K}_{IC}$ for the filled crack.

The table data yield the ultimate strength defined as $\sigma_B = Q_*/(t\pi R) = p_{0*}/(\pi R)$ equal to $\sigma_B \approx 2580$ kPa. Relationship (6.112) gives at $p = p_*$ $K_{IC} = 330$ kPa m$^{1/2}$. The value of $\tilde{K}_{IC}$ follows from Eq. (6.119) at substituting $\beta = l/b = 10$, $\omega = E_1/E = 0.03$, and p $= 294$ kN/m. It is equal to $\tilde{K}_{IC} = 328$ kPa m$^{1/2}$.

The values K_{IC} and $\tilde{K}_{IC}$ differ within the test point scattering.

Load-carrying capacity of a cylindrical element is defined as $p_{0*} = \pi\sigma_B R$. The same element with an empty crack can carry without fracture the load p_* as defined

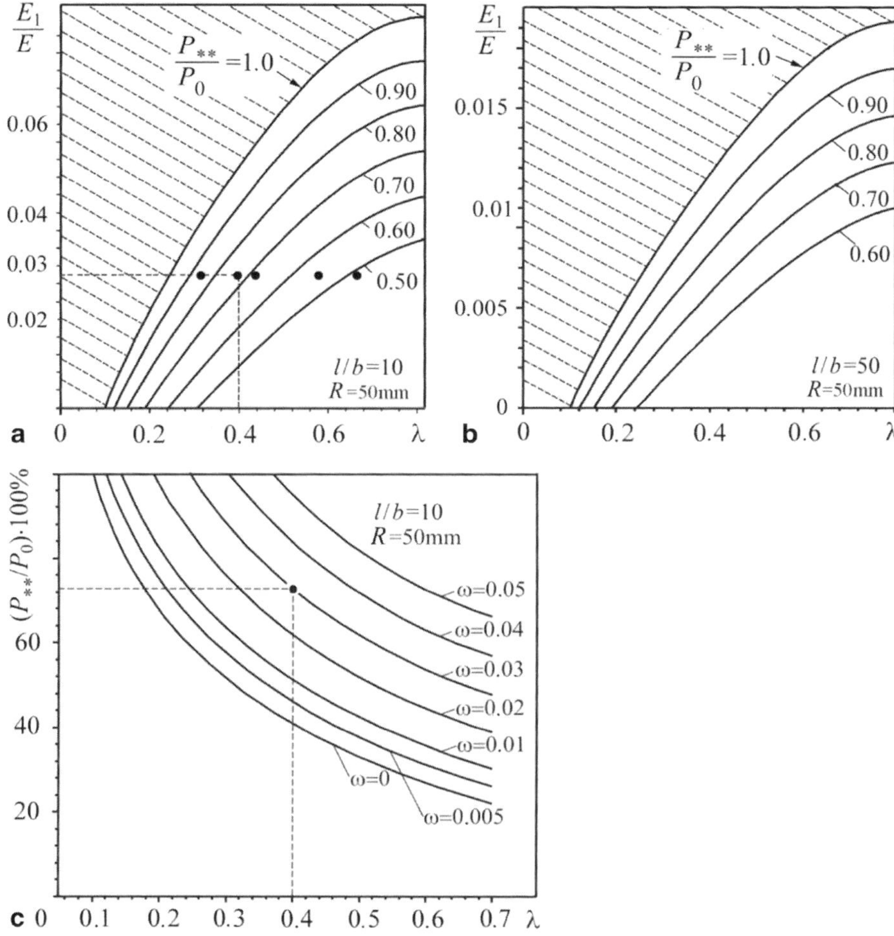

Fig. 6.51 **a, b** Hardness-crack length charts at $\beta = 10$ (**a**) and $\beta = 50$ (**b**) for optimization of injection materials and technologies. Circles present the experimental test data. The hatched area corresponds to a zone of complete renewal. **c** Crack length effect on efficiency of serviceability restoration

by Eq. (6.113). The cylindrical element with a filled crack has the load-carrying capability defined by Eq. (6.120).

Let us now establish parameters of the injection material able to ensure a high efficiency of renewal of a damaged concrete structure. For this purpose, plot values p_{**}/p_0 as functions of $\lambda = l/R$ at various values of $\omega = E_1/E$ and fixed values $\beta = l/b$ using the formula (6.120). Figure 6.51 shows such charts for the cylindrical structural elements (disks) land enables the estimation of the efficiency (level) of renewal of damaged concrete structures using specific injection technologies and/or specific injection compositions (characterized by parameters E_1/E, l/b).

The above-presented results indicate that the strength or serviceability of a damaged concrete structure is renewable through the injection into defects of the material with a hardness (Young modulus) much lower than that of its host material. This conclusion found experimental evidence of concrete element restoration through injection into crack-like defects of the polyurethane compositions with a hardness about 30 times lower than the hardness of the concrete. Indeed, the experimentally established strength of the concrete was restored by about 73 % at $\beta = 10$ and $l/R = 0.4$. Consequently, in this case, the experimental data and the theoretical estimate were in good agreement (Fig. 6.51c).

6.12   Limit Equilibrium of Bodies with Voids and Cracks in the Void Surface[2]

The emergence of crack-like defects in close vicinity to design stress concentrators is a serious problem in engineering practice, giving rise to a demand for serviceability estimations in similar cases. Let us consider some mathematical solutions to such problems.

Tensile Strength of Bodies with Voids and Cracks in the Void Surface

The tensile strength of a deformable body with circular (cylindrical) holes and radial cracks outgoing from its surface (Fig. 6.52) was subject to study through various techniques [2], [31–34].

The particular solution for ultimate load $p = p_c$ initiating the fracture has form [2]:

$$p_c = \frac{K_{IC}\sqrt{\pi(b+a)}}{f_1(a,b,R)},\tag{6.122}$$

$$f_1(a,b,R) = A(a,b,R)\sqrt{(a+R)(b-R)} - A(a,b,-R)\sqrt{(a-R)(b+R)}$$

$$-A(a,b,R)\ln\frac{[\sqrt{ab}+\sqrt{(a-R)(b+R)}]^2 + R^2}{[\sqrt{ab}+\sqrt{(a+R)(b-R)}]^2 + R^2}$$

$$+\frac{a+b}{2}\left(\pi + \arcsin\frac{a-b-2R}{a+b} - \arcsin\frac{a-b+2R}{a+b}\right);\tag{6.123}$$

$$A(a,b,\pm R) = 1 \pm \frac{R}{b} + \frac{R^2}{b^2}\left(\frac{5}{8} + \frac{1}{8}\frac{b}{a}\right) \pm \frac{R^3}{b^3}\left(\frac{15}{16} + \frac{1}{4}\frac{b}{a} - \frac{3}{16}\frac{b^2}{a^2}\right);$$

$$B(a,b,R) = \frac{R^2\sqrt{ab}}{32a^3b^4}[8a^2b^2(a+b) + 3R^2(7a^3 - 3a^2b + 5ab^2 - b^3)];$$

$$a = R + l_1; b = R + l_2.$$

[2] The present section contains results obtained with the assistance of N.V. Onischak.

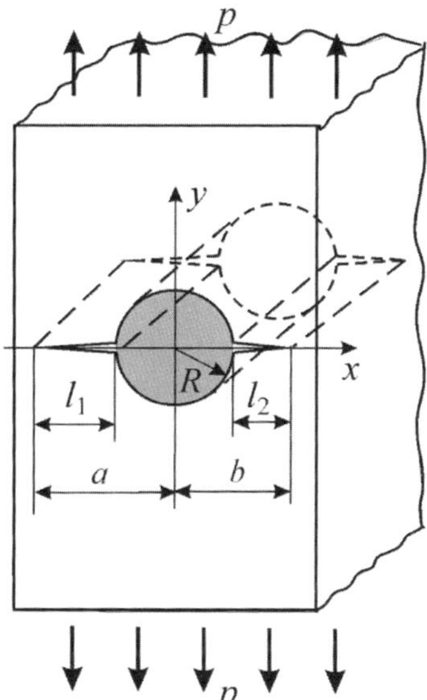

Fig. 6.52 Tension of a body with a circular hole and cracks outgoing from its surface

In the case of $l_1 = l_2 = l$ (Bowie problem [31]), Eq. (6.122) becomes essentially simpler:

$$p_c = \frac{K_{IC}\sqrt{\pi}}{2\sqrt{R(1+\xi)}} \cdot \frac{1}{f_2(\xi)},$$

$$f_2(\xi) = \frac{\pi}{2} - \arctan\frac{1}{\sqrt{2\xi+\xi^2}} + \frac{\sqrt{2\xi+\xi^2}}{(1+\xi)^4}(2+2\xi+\xi^2), \quad \xi = \frac{l}{R}.$$

$$(6.124)$$

In the case of an infinite plate weakened by a hole with a single crack, i.e., $l_2 = 0$ in Eqs. (6.122) and (6.123), the ultimate load is equal to [2]:

$$p_c = \frac{K_{IC}\sqrt{\pi}}{2\sqrt{R(1+\xi_1)}} \cdot \frac{1}{f_3(\lambda)},$$

$$f_3(\lambda) = A(\lambda)\sqrt{2\lambda(1-\lambda)} - B(\lambda)\ln\frac{1+\lambda}{[1+\sqrt{2(1-\lambda)}]^2 + \lambda}$$

$$+\frac{1+\lambda}{2}\left(\frac{\pi}{2} + \arcsin\frac{1-3\lambda}{1+\lambda}\right),$$

$$A(\lambda) = \frac{1}{16}(16 + 15\lambda + 14\lambda^2 + 15\lambda^3),$$

$$B(\lambda) = \frac{\lambda\sqrt{\lambda}}{32}(5 + 23\lambda - 9\lambda^2 + 21\lambda^3),$$

$$\lambda = \frac{R}{b} = \frac{1}{1+\xi_1}, \xi_1 = \frac{l_1}{R}, b = R + l_2.$$

$$(6.125)$$

Fig. 6.53 Tension of a body
with a filled hole and cracks

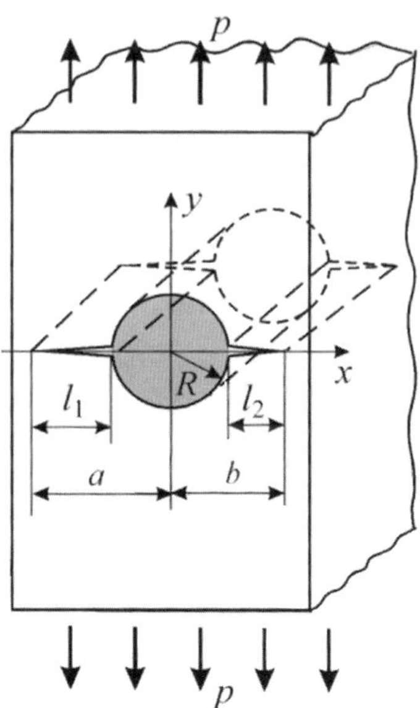

Let us now establish how this will change the strength of the body due to filling of
the hole with the injection material (Fig. 6.53). For this purpose, let us consider the
respective two-dimensional problem of elastic theory. In the given case, stresses in
the plane $y = 0$ have the appearance:

$$\sigma_{yy}(x,0) = \sigma_{yy}^0(x,0) + \sigma_{yy}^1(x,0), \tag{6.126}$$

where $\sigma_{yy}^0(x,0)$ are stresses arising in the plate with a filled hole but without a crack
under tension by applied force intensity p; $\sigma_{yy}^1(x,0)$ are stresses arising along the
crack within the range $-a \leq x \leq b$ due to the following forces applied to the crack
surfaces:

$$q_n(x) = \begin{cases} -\sigma_{yy}^0(x,0), -a \leq x \leq -R \\ 0, -R < x < R \\ -\sigma_{yy}^0(x,0), R \leq x \leq b \end{cases} \tag{6.127}$$

Based on results outlined in [23], we have the following formula to determine the
stresses $\sigma_{yy}^0(x,0)$:

$$\sigma_{yy}^0(x,0) = p\left(1 + \frac{1}{2}\delta_1\frac{R^2}{x^2} - \frac{3}{2}\delta_2\frac{R^4}{x^4}\right), \delta_1 = \frac{\mu(\kappa_1 - 1) - \mu_1(\kappa - 1)}{2\mu_1 + \mu(\kappa_1 - 1)},$$

$$\delta_2 = \frac{\mu_1 - \mu}{\mu + \mu_1\kappa}, \quad \kappa = 3 - 4v, \quad \kappa_1 = 3 - 4v_1. \tag{6.128}$$

Here, μ, μ_1, ν, and ν_1 are the shear moduli and Poisson ratios of the host and guest materials, respectively. In turn, to define stresses $\sigma_{yy}^1(x,0)$, the formula [2] is valid:

$$\sigma_{yy}^{(1)}(x,0) = \frac{1}{\pi\sqrt{(x-b)(x+a)}} \int\limits_{-a}^{b} \frac{q_n(t)\sqrt{(t-b)(t+a)}}{|a-t|} dt. \tag{6.129}$$

Then, the ultimate force p_c follows from the criterion condition $K_1(p,a,b) = K_{\mathrm{IC}}$. In other words, we have initially to perform the necessary calculations using formulae (6.128) and (6.129) in order to find stresses $\sigma_{yy}^0(x,0)$ and $\sigma_{yy}^1(x,0)$, and, finally, determine stresses $\sigma_{yy}(x,0)$ and intensity factor $K_1(p,a,b)$ using formula (6.126):

$$p_c = \frac{\sqrt{a+b}}{f(a,b,R)} K_{\mathrm{IC}},$$

$$f(a,b,R) = \left(\int\limits_{-a}^{-R} + \int\limits_{R}^{b}\right)\left(1 + \frac{1}{2}\delta_1\frac{R^2}{x^2} - \frac{3}{2}\delta_2\frac{R^4}{x^4}\right)\frac{\sqrt{a+x}}{\sqrt{b-x}} dx. \tag{6.130}$$

For cracks of identical length ($b=a$) we have:

$$f(a,R) = 2a\left[\frac{\pi}{2} - \arcsin\frac{R}{a} + \delta_1\frac{R\sqrt{a^2-R^2}}{2a^2}\right.$$
$$\left. -\delta_2\frac{R(2a^2+4R^2+3aR)\sqrt{a^2-R^2}}{2a^4}\right]. \tag{6.131}$$

As a result, the following expression for the ultimate load $p = p_c$ is derivable from Eqs. (6.12) and (6.130) combined with criterion $K_1(a,p_c) = K_{\mathrm{IC}}$:

$$p_c = K_{\mathrm{IC}}\left\{\sqrt{2a}\left[\frac{\pi}{2} - \arcsin\frac{R}{a} + \delta_1\frac{R\sqrt{a^2-R^2}}{2a^2} - \delta_2\frac{R(2a^2+4R^2+3aR)\sqrt{a^2-R^2}}{2a^4}\right]\right\}^{-1}. \tag{6.132}$$

Figure 6.54 graphically presents dependence of the ultimate load p_c on crack length and the hardness of the injection material calculated using Eq. (6.132).

One can see from the figure that the serviceability of the structures with holes and outgoing cracks is restored efficiently only along small crack lengths ($l < 0.5R$). Along longer cracks, they are subject to separate injection procedures.

Compressive Strength of Bodies with Voids and Cracks in the Void Surface

The key role of holes and voids in crack nucleation in a concrete structure is well known. Even under compression, the concentration of tensile stress arises near holes

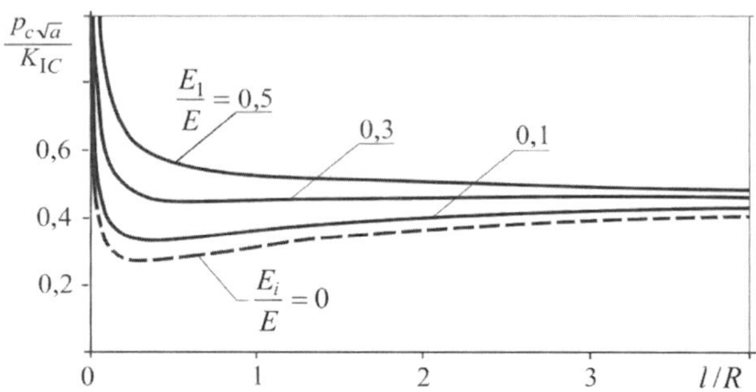

Fig. 6.54 Strength of a body with a filled hole and two identical cracks at various values of filler hardness

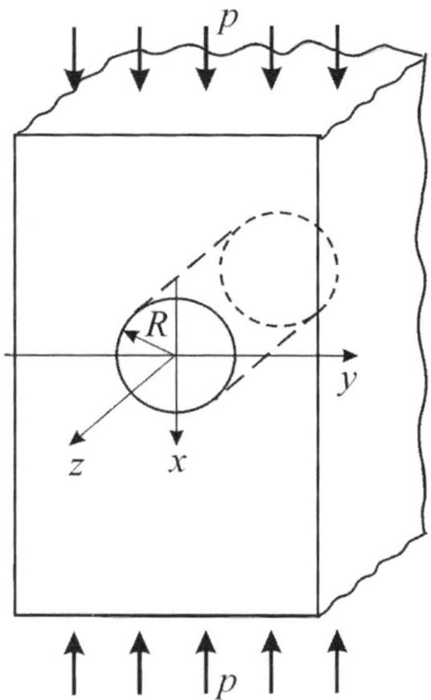

Fig. 6.55 Body with circular hole under compression

or voids facilitating crack initiation and propagation [35]. According to published solutions of elastic theory problems [23], a zone of tensile stresses arises near a circular hole in a large body under compression (Fig. 6.55) in the plane $y = 0$ with magnitude determined by the relationship:

$$\sigma_{yy}(x,0) = -p\left(\frac{1}{2}\frac{R^2}{x^2} - \frac{3}{2}\frac{R^4}{x^4}\right),\tag{6.133}$$

where p is the intensity of compressive force and R is the hole radius.

Figure 6.56 shows stress distribution around the empty void ($E_1/E = 0$). One can see that tensile stresses arise in the range $R < x < 1.7R$ while at $x > 1.7R$ they change into quite small compressive stresses.

At a high enough intensity of compressive forces, p tensile stresses in plane $y = 0$ within the range $R < x < 1.7R$ can initiate cleavage crack nucleation near the hole.

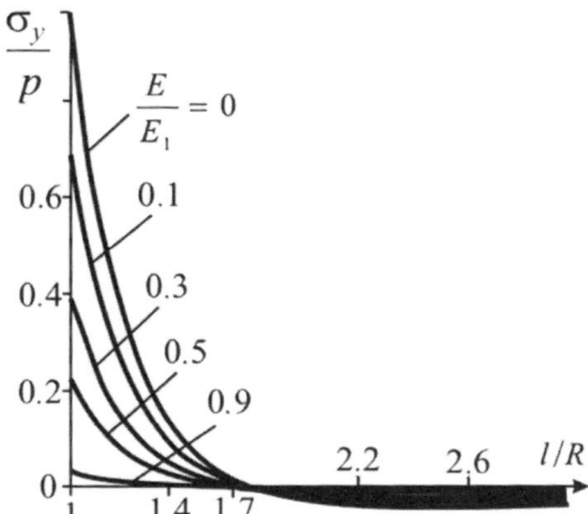

Fig. 6.56 Stress distribution beyond the filled hole at various values of filler hardness

We have studied experimentally the compressive crushing of prismatic specimens $100 \times 200 \times 300$ mm in size made from sand-cement mortar with through holes 20 mm in diameter along the prism axis at a mid-height position (Fig. 6.57). The obtained results confirm crack nucleation and propagation in the above-indicated plane. Similar results are available in other references as well [35], [36].

Crack nucleation near the hole was observable at compressive loads close to 50 % of the ultimate value. At further loading, crack length increased up to a certain value and then the crack became stable since its tip entered into the zone of compression.

Final fracture of specimens with a hole took place at the same loads as that of the fracture of specimens without holes. The main crack path in all cases intercepted the hole (Fig. 6.57a). Such fracture behavior may have the following explanation. It follows from Eq. (6.133) that at $x = R$, stress concentration near the hole is independent of the hole radius. Therefore, inevitably present in concrete, voids and artificial holes generate a certain stress concentration in the material. These defects initiate crack nucleation. However, when such cracks enter into the zone of compression (see Fig. 6.56), they become stable at the applied load values they nucleated. Subsequent crack growth proceeds in accordance with laws of fracture mechanics. Occurring in the body, cracks grow, merge and form the main crack at further load increase or long-term action. The main crack finally breaks down the body.

One conclusion from the above is that diminishing the concrete porosity in the stage of preparation through existing technological methods such as vibration, water proportion optimization, etc., is the efficient way to strengthen concrete structures. Nevertheless, complete suppression of pore formation is hardly attainable. In this view, the construction practice relies on strengthening of the concrete structural elements by means of filling the pores or voids with liquid materials able to polymerize or crystallize after a certain period. In other words, impregnation of concrete by various solutions takes place. In particular, voids and cracks are removable through the pumping of aqueous cement mortars or polymer materials, for instance, polyurethanes.

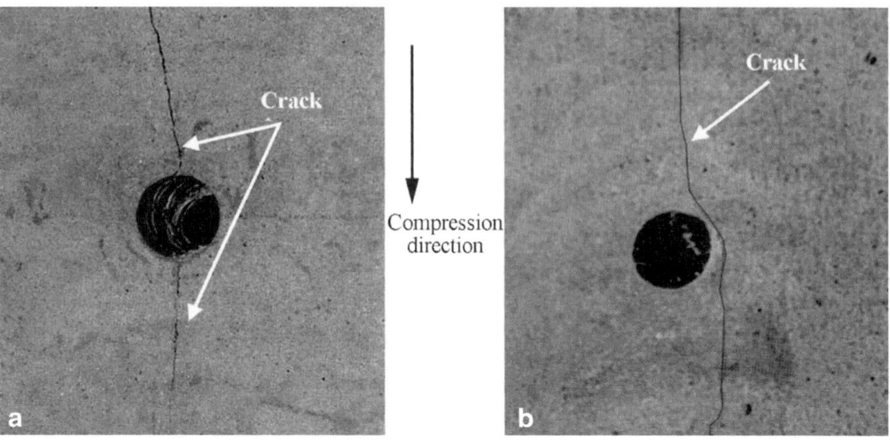

Fig. 6.57 Fragments of a prismatic specimen surface with an empty hole (**a**) or a filled hole (**b**) after fracture

In the assumption of ideal mechanical contact at the interface of materials after solidification, we can write stresses near the filled hole based on solutions of elastic theory [23] as:

$$\sigma_{yy}^{0}(x, 0) = -p \left(\frac{1}{2}\delta_1 \frac{R^2}{x^2} + \frac{3}{2}\delta_2 \frac{R^4}{x^4} \right), \tag{6.134}$$

where

$$\delta_1 = \frac{\mu(\kappa_1 - 1) - \mu_1(\kappa - 1)}{2\mu_1 + \mu(\kappa_1 - 1)}, \delta_2 = \frac{\mu_1 - \mu}{\mu + \mu_1\kappa}, \mu = \frac{E}{2(1 + \nu)},$$

$$\mu_1 = \frac{E_1}{2(1 + \nu_1)}, \kappa = 3 - 4\nu, \kappa_1 = 3 - 4\nu_1,$$

E, E_1, ν, ν_1 are the Young moduli and Poisson ratios of the host and guest materials, respectively.

Figure 6.56 shows plots of the function (6.134) at various values of filler hardness. We can see that essential tensile stress relief takes place at filler hardness values closest to the hardness of the matrix. Nevertheless, even material with relative hardness $E_1/E = 0.1$ decreases tensile stress concentration near the hole by 30 %. As a result, we can expect changes in the process of main crack formation under compression.

The dominant mechanism of this process consists of microcrack nucleation and merging into a macrocrack penetrating the concrete near the hole. When the hole is empty, the process runs at lower applied forces than when the hole is filled. Experimental studies of prismatic specimens with filled holes, as well as Eqs. (6.133) and (6.134), confirm this conclusion.

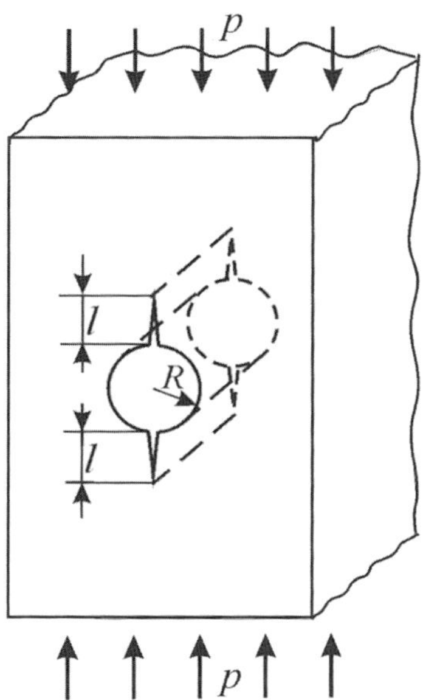

Fig. 6.58 Body with a circular cylindrical hole and cracks under compression

Fracture Criterion for Cracks in a Circular Hole Surface

Using criteria of fracture mechanics and the mathematical tools of crack theory outlined in [2], one can easily write the ultimate compressive load $p = p_c$ corresponding to the limit equilibrium state for cracks formed in the contour of a circular hole (Fig. 6.58). This load has the appearance [35]:

$$p_c = K_{\mathrm{IC}}\sqrt{\frac{\pi(1+\lambda)^7}{4R[(1+\lambda)^2 - 1]}}, \quad \lambda \equiv \frac{l}{R}, \tag{6.135}$$

where K_{IC} is the fracture toughness of the material.

It is interesting that, in such a loading configuration, the function $p_c(l)$ has a minimum at the certain point $l = l_*$ ($l_* \approx 0.18R$). At $l < l_*$ the crack propagates catastrophically while at $l > l_*$ the crack growth is stable, i.e., crack length increment requires load increment. Such behavior of the function $p_c(l)$ is understandable in terms of stress distribution around the hole, which is characterized by the tension zone changing into a compression zone at $l > 0.73R$. Taking into consideration the above-mentioned, let us consider the limit equilibrium state in a compressed body with the circular hole filled by a foreign material.

We shall assume that, after solidification of the liquid paste introduced into the hole, the bonds form in the host (concrete)-guest interface, ensuring the perfect mechanical contact. However, we consider the cracks empty after such a procedure.

The linear crack theory approximately describes this situation as a boundary problem for an infinite body containing a mathematical slit with length $2l$ and forces applied to its edges equal to

$$\sigma_{yy}(x) = \sigma_p(x) = \begin{cases} -\sigma_{yy}^0(x), -l \le x < -R, \\ 0, -R \le x \le R, \\ -\sigma_{yy}^0(x), R < x \le l. \end{cases} \tag{6.136}$$

Here $\sigma_{yy}^0(x)$ are stresses defined by Eq. (6.134).

The boundary problem (6.136) for the mathematical slit easily transforms (see, e.g., [23]) to a singular integral equation

$$\int_{-l}^{l} \frac{u'(t)dt}{t-x} = \frac{2\pi(1-v^2)}{E}\sigma_p(x), \quad -l \le x \le l, \tag{6.137}$$

where $u'(t)$ is displacement of the slit edges within the range $[-l, l]$. Eq. (6.137) has the exact solution [23]:

$$u'(x) = -\frac{2(1-v^2)}{\pi E \sqrt{l^2-x^2}} \int_{-l}^{l} \frac{\sqrt{l^2-t^2}\sigma_p(t)}{t-x}dt. \tag{6.138}$$

Considering SIF definition in the form $K_I = E/[2(1-v^2)]$ $\lim_{x \to \pm l} \sqrt{\pi(l^2-x^2)/l}u'(x)$, one can find:

$$K_1(p,l,R) = -\frac{p\sqrt{l}}{\sqrt{\pi}}\left(\delta_1 \frac{R\sqrt{l^2-R^2}}{l^2} + \delta_2 \frac{R(2l^2+4R^2+3lR)\sqrt{l^2-R^2}}{l^4}\right).$$

Combining the expression for K_I and criterion $K_I(p_c, l, R) = K_I$, we obtain

$$p_c = -\frac{K_{1C}\sqrt{\pi}}{\sqrt{l}\left(\delta_1 \frac{R\sqrt{l^2-R^2}}{l^2} + \delta_2 \frac{R(2l^2+4R^2+3lR)\sqrt{l^2-R^2}}{l^4}\right)} \tag{6.139}$$

Figure 6.59 graphically presents dependence of the ultimate load p_c on crack length at various values of the injection material's hardness. These results allow for the conclusion that effective strengthening of a body with a hole and cracks is possible only when using a sufficiently hard filler. The above feature discriminates between holes and cracks at injection since the latter demonstrate a strong strengthening effect even at minimal filler hardness [30]. The point $l = l_*$ ($l_* \approx 1.18R$) is of a special interest since the function $p_c(l)$ reaches a minimum at this point. At $l < l_*$ the crack propagates catastrophically while at $l > l_*$ the crack growth is stable, i.e., crack length increment requires load increment.

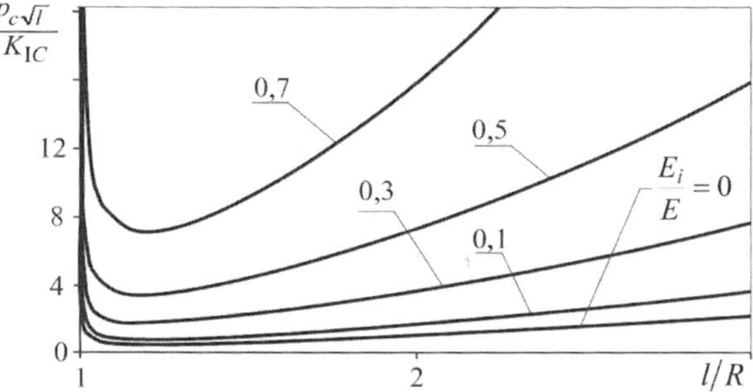

Fig. 6.59 Strength of a body with a filled hole and cracks (Fig. 6.58) at various values of filler hardness (E_1/E)

In conclusion, the filling of pores and voids in concrete structural elements working under compression through liquid materials capable of polymerization or crystallization in a certain time can be the most effective method for strengthening such elements. The harder the filler after solidification, the more effective such strengthening will be.

References

1. Czarnecki L, Emmons PH (2002) Naprava i ochrona konstrukcji betonowych (Repair and protection of concrete structures. Polski Cement, Krakov
2. Panasyuk VV (1968) Predel'noye ravnovesiye khrupkikh tel s treshchinami (The limiting equilibrium of brittle bodies with cracks). Nauk. dumka, Kyiv
3. Cherepanov GP (1977) Mekhanika khrupkogo razrusheniya (Mechanics of brittle fracture). Stroyizdat: Moscow
4. Panasyuk VV (1991) Mehanika kvazikhrupkogo razrusheniya materialov (Mechanics of quasi-brittle fracture of materials). Nauk. dumka, Kyiv
5. Liebowitz H (ed) (1968) Mathematical fundamentals. In: Freudenthal AM (ed) Fracture, an advanced treatise, vol 2. Academic Press, New York
6. Panasyuk VV, Stadnik MM, Sylovanyuk VP (1986) Kontsentratsiya napyazhenii v trekhmernykh telakh s tonkimi vklyucheniyami (Stress concentration in three-dimensional solids with thin inclusions). Nauk. dumka, Kyiv
7. Popov GY (1982) Kontsentratsiya uprugikh napyazhenii vozle shtampov, razrezov, tonkikh vklyuchenii i podkreplenii (Elastic stress concentration near punches, slits, thin inclusions, and reinforcements). Stroyizdat, Moscow
8. Alexandrov VM, Mkhitaryan SM (1983) Kontaktnyye zadachi dlya tel s tonkimi pokrytiyami i prosloikami (Contact problems for bodies with thin coverings and interlayers). Stroyizdat, Moscow
9. Sylovanyuk VP (2000) Ruinuvannya poperedn'o napruzhenykh i transversalno-izotropnykh til iz defektami (Fracture of prestress with defects). Karpenko Phys. Mech. Inst. NASU, Lviv
10. Sulim GT (2007) Osnovy matematichnoi teorii termopruzhnoi rivnovagi deformivnykh tverdykh til z tonkimi vklyucheniyami (Mathematical fundamentals of thermo elastic equilibrium in deformable solids with thin inclusions). Res. & Ed. Center, Lviv
11. Griffith AA (1920) The phenomenon of rupture and flow in solids. Phil Trans Roy Soc Ser A 221:163–198
12. Irwin GR (1957) Analysis of stresses and strains near the end of crack traversing a plate. J Appl Mech 24:361–364

13. Panasyuk VV, Berezhnitskii LT (1964) Ultimate load determination at tension of plate with bow-shaped crack. Probl Mech Real Solids 3:3–19
14. Panasyuk VV, Berezhnitskii LT, Kovchik SE (1964) On propagation of arbitrarily oriented straight crack. Appl Mech 1:48–55
15. Panasyuk VV, Andreikiv AE, Kovchik SE (1977) Metody otsenki treshchinostoykosti konstruktsionnykh materialov (Methods of fracture toughness estimation in structural materials). Nauk. dumka, Kyiv
16. Leonov MY, Panasyuk VV (1959) Small crack growth in solids. Appl Mech 5:391–401
17. Panasyuk VV (1960) On the theory of crack propagation during deformation of brittle solids. Rep Acad Sci Ukr SSR 9:1183–1185
18. Vitvitskii PM, Leonov MY (1961) On fracture of plate with slit. Appl Mech 7:516–520
19. Dugdale DS (1960) Yielding of steel sheets containing slits. J Mech Phys Solids 8:100–108
20. Wells AA (1961) Critical tip opening displacement as fracture criterion. In: Proc. Crack Propagation Symp., Cranfield, vol 1, pp 210–221
21. Andreikiv AE (1979) Razrusheniye kvazikhrupkikh tel s treshchinami pri slozhnom napryazhennom sostoyanii (Fracture of quasi-brittle solids with cracks under complex stress state). Nauk. dumka, Kyiv
22. Panasyuk VV, Savruk MP, Datsyshin AP (1976) Raspredeleniye napyazhenii okolo treshchin v plastinakh i obolochkakh (Stress distribution around cracks in plates and shells). Nauk. dumka, Kyiv
23. Mushelishvili NI (1966) Nekototyye osnovnyye zadachi matematicheskoi teorii uprugosti (Selected general problems of mathematical elastic theory). Stroyizdat: Moscow
24. Yarema SY, Krestin GS (1966) Cohesion modulus determination in brittle materials by means of compressive testing of disks with cracks. Phys Chem Mech Mater 1:10–14
25. Sylovanyuk VP, Yukhim RY (2007) Deformation and fracture of materials near inclusions under static loading of body. Phys Chem Mech Mater 6:31–35
26. Sylovanyuk VP, Marukha VI, Onishchak NV (2009) Strength of body with crack partially filled with injection material. Phys Chem Mech Mater 5:77–87
27. Cherepanov GP (1983) Mehanika razrusheniya kompozitsionnykh materialov (Fracture mechanics of composite materials). Stroyizdat, Moscow
28. Onishchak NV (2009) Otsinki zmitsnennya tila z dvoma trishchinami, "zalikovannymi" za inyektsionnymi tehnologiyami (Strengthening estimations for body with two cracks healed using inlection technologies). In: Panasyuk VV (ed) Mekhanika ruinuvannya materialiv i mitsnist' konstruktsii (Fracture mechanics of materials and strength of structures). Karpenko Phys. Mech. Inst. NASU, Lviv, pp 561–564
29. Onishchak NV (2008) Mitsnist' tila iz poverkhnevoyu trishchinoyu, zapovnenoyu pruzhnym materialom (Strength of body with surface crack filled by elastic material). In: Suchasni problemy mekhaniki ta matematiki (Modern problems in mechanics and mathematics), vol 2, Lviv, pp 80–82
30. Sylovanyuk VP, Marukha VI, Onishchak NV (2007) Residual strength of cylindrical elements with cracks healed using injection technologies. Phys Chem Mech Mater 1:99–104
31. Bowie OL (1956) Analysis of infinite plate containing cracks, originating at the boundary of an internal circular hole. J Math Phys 35:60–71
32. Berezhnitskii LT (1966) On propagation of cracks outgoing from contour of curvilinear hole in plate. Phys Chem Mech Mater 1:21–31
33. Kaminskii AO (1964) Brittle fracture of plate weakened by curvilinear hole with cracks. Appl Mech 10:375–381
34. Panasyuk VV (1965) On the fracture stress for plate weakened by circular hole with radial cracks. Rep Acad Sci Ukr SSR 7:868–871
35. Zaitsev YV (1982) Modelirovaniye deformatsiy i prochnosti betona metodami mekhaniki razrusheniya (The concrete deformations and strength simulation using fracture mechanics methods). Stroyizdat, Moscow
36. Raju NK (1971) Strain distribution and microcracking in concrete prisms with a circular hole under uniaxial compression. J Mater 5:450–456

Chapter 7
Methods and Devices for Technical Diagnostics of Long-term Concrete Structures

Abstract The analysis of available publications and patents related to methods and means (devices) for non-destructive testing and quantitative status estimation of long-term concrete and reinforced concrete structures damaged by cracks and/or other dangerous defects is given. The authors provide and characterize a classification scheme for non-destructive testing methods used to determine initial mechanical parameters of concrete and reinforced concrete building structures and those degraded during operation. They consider and evaluate the conventional in building practice techniques and monitoring devices for technical diagnostics of damaged concrete structures. They also analyze the main physical and mechanical methods for laboratory testing of concrete and reinforced concrete specimens as well as methods for technical diagnostics of the status of structures of long-term service. The presented methods include visual inspection, optical, acoustic (or ultrasound), and radiation techniques and devices for crack and/or defect detection and sizing. Description of electrochemical techniques and devices for measuring the extent of degradation in steel concrete reinforcing elements with the aim of making decisions as to the possibility of further reliable use of the concrete and reinforced concrete structures damaged by cracks and/or other defects is provided. Case studies of use of up-to-date technical diagnostic devices for such purposes are considered. The design of a computerized truck-mounted mobile diagnostic-restoration complex for major repair of the sewage collectors and main pipelines is developed. Operational parameters of diagnostic or processing devices, appliances and systems of the complex are included. Practical implementation of the complex in industrial and municipal building structures is illustrated.

Technical diagnostics of structural elements, including concrete elements, use various non-destructive testing methods. These methods provide data on the change of material characteristics in the course of service life, various types of damages or injuries, crack-like defects, etc. [1]. Acute crack-like voids are the most dangerous among possible common defects in structural elements of bridges, dams, and other structures [2]. For this reason, these defects alone attract the most attention during development and improvement of accuracy of inspection techniques and residual resource estimation methods for damaged objects [3], as well as effective technologies for the restoration of serviceability [4], [5].

V. V. Panasyuk et al., *Injection Technologies for the Repair of Damaged Concrete Structures,* DOI 10.1007/978-94-007-7908-2_7,
© Springer Science+Business Media Dordrecht 2014

The most frequent cause of damaging of concrete, reinforced concrete, and brick structural elements is the interaction of such materials with reactive environments. Corrosion and service degradation of construction materials in structures of long-term operation are very versatile in mechanism and appearance. The conventional physical methods for determining the defective status of materials and structures according to Ukrainian standards [6] are available from the reference books on non-destructive testing and diagnostic tools and methods [2]. However, simple but convenient mechanical tools and methods can also be used for diagnostics of building materials and structures.

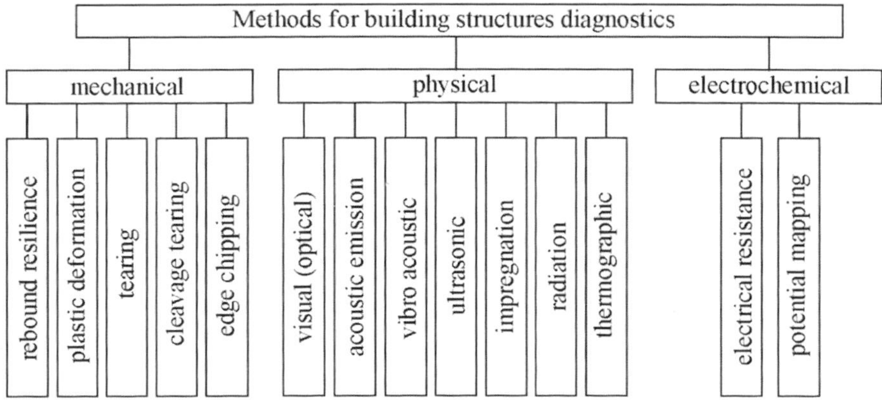

The table above presents types of non-destructive testing methods for checking the mechanical characteristics of building materials and structures. It is provisional but informative as to existing modifications.

7.1   Mechanical Tools and Methods of Technical Diagnostics

The mechanical methods of diagnostics are most widespread since they require no complicated electronic equipment and are available to use at any structure site.

Method of Elastic Impact Response (Rebound Resilience)

The Schmidt hammer (Fig. 7.1) is one of most common tools in engineering practice for non-destructive testing of concretes. The basis of Schmidt hammer application consists of the relationship between the compressive strength of concrete (R_b) and the height of the loading head rebound from the striking indenter tightly pressed to the concrete surface. According to USSR standard GOST 22690.1-77, the loading head must have a kinetic energy over 0.75 J while the radius of the indenter's spherical tip must be at least 5 mm.

At present, a number of devices exist designed for measuring the head rebound height or distance that make it possible, as distinct from the first models of Schmidt

Fig. 7.1 One of first versions
of a Schmidt hammer for
horizontal surfaces. [7]

sclerometer (Fig. 7.1), to inspect vertical, overhead, inclined surfaces and, above all, hard-to-reach corners of building structures (Fig. 7.2) [7].

However, sclerometer hammers of all modifications possess one general disadvantage connected with the necessity of plotting dependences of the concrete's strength on the head rebound height for each concrete grade (see Sect. 7.3). Schmidt sclerometers of type P (Fig. 7.2) are suitable for objective estimations of the concrete grade in a long-term building or structure but are lacking in obtaining quantitative estimations of compressive strength for aged concrete, which are quite problematic.

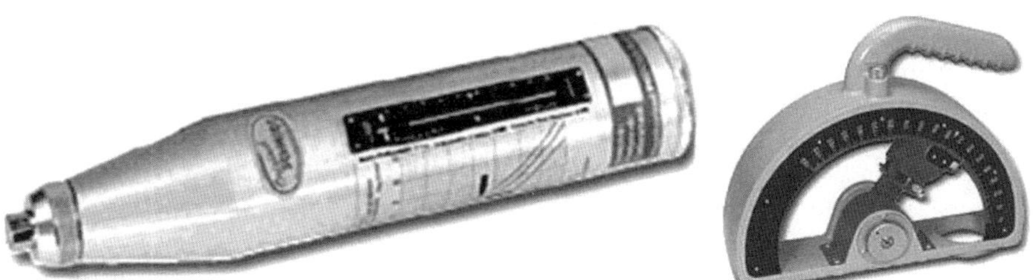

Fig. 7.2 Schmidt sclerometers: type P (*left*) and type N (*right*)

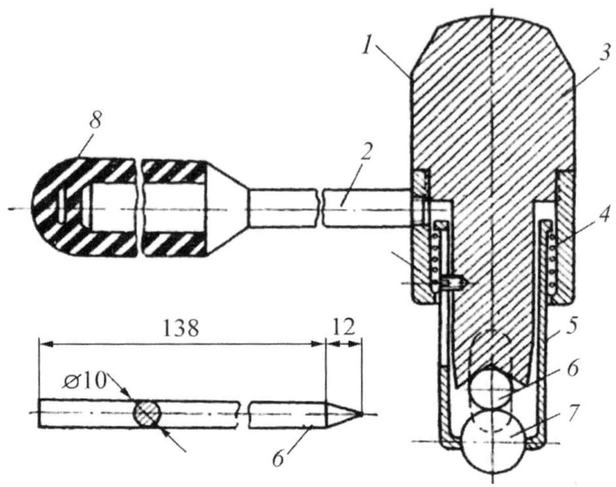

Fig. 7.3 Design of standard Kashkarov hammer including: case (*1*); handle (*2*); head (*3*); spring (*4*); cylinder (*5*); reference rod (*6*); ball (*7*); knob (*8*)

Method of Plastic Deformation

At the constant kinetic energy of a striking head, indent size characterizes the hardness of the concrete surface layer.

The indent formed due to the indenter strike has a characteristic diameter d and depth h. The indent diameter must fall within the range $0.3D < d < 0.7D$, where D is the indenter diameter. Some device modifications are also available using intenders in the shape of disks [7].

Concrete strength in the structure results from the size-compressive strength of the calibration curves indent plotted using tests of reference specimens. However, this method is indirect as well.

In this view, the standard Kashkarov hammer deserves attention (USSR standard GOST 22690.2-77) operating with reference metal rods with a known specified hardness [8]. The hammer has a head with a steel ball 15.88 mm in diameter. The head case has a cavity for inserting the reference rod between the ball and anvil. The reference rod 12 or 10 mm in diameter is subject to manufacturing from mild steel VSt3sp2 with ultimate strength $\sigma_B = 420 \dots 460$ MPa [7], [8]. The procedure of concrete strength determination is as follows (Fig. 7.3).

The operator either strikes a concrete surface with the hammer or puts it to the selected point and strikes the head *3* with a bench hammer. The ball produces indents simultaneously in the concrete and the reference rod. After each strike, the operator shifts the reference rod in the head case by $10 \dots 12$ mm in order to obtain a new indent. The operator measures the ball indent diameters in concrete d_σ and reference rod d_e using a measuring magnifier. Then, he calculates the average indent ratio:

$$\left(\frac{d_\sigma}{d_e}\right)_{ave} = \frac{1}{N}\sum_{i=1}^{N}\frac{d_{\sigma i}}{d_{ei}} \tag{7.1}$$

Fig. 7.4 Method of tearing

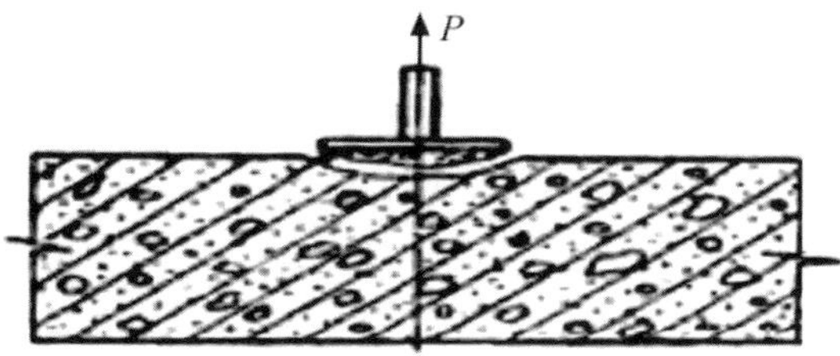

and determines the concrete strength using the calibration curve $R_b = f(d_\sigma/d_e)$. The steel rod in the hammer enables him to compare the strength values for various concrete brands and grades, which is of primary importance in the diagnostics of buildings or structures under long-term operation.

Method of Tearing

The concrete tearing strength according to GOST 22690.3-77 has to be measured using a tearing off steel disk glued onto the concrete surface (Fig. 7.4)

The steel disk with a threaded stem on one side is subject to gluing onto a pre-trimmed concrete surface. After curing the glue, the operator tears the disk off using an appliance capable of measuring the tearing force, for example, GPNV-5 [7]. During the test, the adjoining concrete area breaks off together with the disk. The division of the tearing force by the disk area, in neglecting the difference between this area and the actual tearing surface area, yields the concrete's strength (Fig. 7.4). The compressive concrete strength value is available from a calibration curve compressive R_b vs. tensile R_{bt} strength.

The company Proceq SA (Switzerland) has developed an advanced method for testing concrete by tearing [9]. The method includes cutting a groove down to a specified depth around the steel disk using an annular cylindrical tool. Thereafter, a special appliance DYNA Z6 or DYNA Z16 (Fig. 7.5a), with the same functional

Fig. 7.5 The device DYNA for measuring concrete tearing strength (**a**), and annular diamond bit for forming a groove around the disk (**b**)

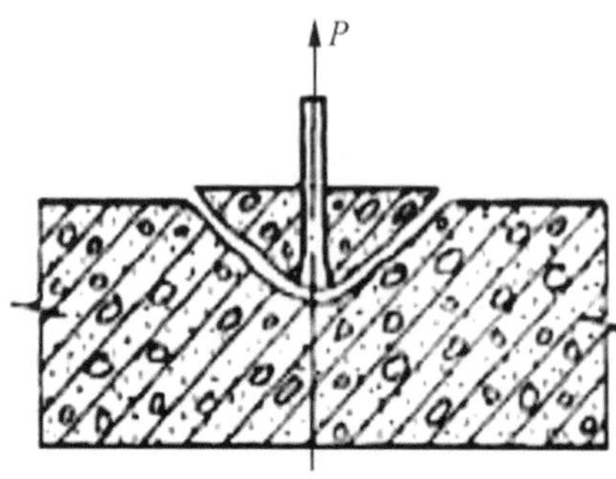

Fig. 7.6 Method of cleavage tearing

concept as the usual adhesion meter, tears off the concrete core with a diameter equal to the diameter of a steel disk so formed by a diamond cutter (Fig. 7.5b).

This method makes it possible to trace the change in the strength of the concrete with depth into a structural element wall or a bulk concrete, by means of raising the depth of the annular groove.

The DYNA devices may have either digital indication of measurement results (Fig. 7.5) or a separate indicator block DYNAMETER [9] equipped with memory, LCD display, and software for data processing and transfer to a personal computer. The devices are operable at temperatures from $-10\,°\text{C}$ to $+60\,°\text{C}$.

Method of Cleavage Tearing

According to GOST 21243-75, compressive concrete strength is determined based on the force P of concrete lump cleavage tearing from a structure. A supplementary characteristic of a calibration curve is that strength is basic here just as in other non-destructive testing methods. However, the author [7] believes that the $P - R_b$ relation is closer and less dependent on side effects than in the case of any other supplementary characteristics. Such a feature allows for the application of pre-rated calibration curves without individual determination for each concrete under study. While ensuring higher accuracy and reliability, this method is distinct for its larger labor-intensity.

This method consists of embedding an anchorage into a concrete body and then tearing it off using a special appliance that enables us to measure the tearing force with a specified accuracy. The cleavage implies breaking off the concrete lump together with the anchorage from a concrete mass (Fig. 7.6).

A nomogram $R_b - f(P)$ enables us thereafter to find the concrete's compressive strength. The anchorage (Fig. 7.7) is embedded into a borehole drilled at the specified point, fixed, and torn off with a concrete lump using appliances such as GPNS-4, GPNS-5 or others according to GOST 21243-75 [7], [8].

Proceq SA (Switzerland) proposes [9] an all-purpose appliance in four modifications enabling cleavage tearing tests as well as holding power tests for stems, plugs, and screws in structural elements (Fig. 7.8).

Fig. 7.7 Schematic view of a
self-locking anchorage
suitable for long-term
structure diagnostics
including an anchor bar with
a cone head (*1*) and serrated
segmental sidepieces (*2*). [8]

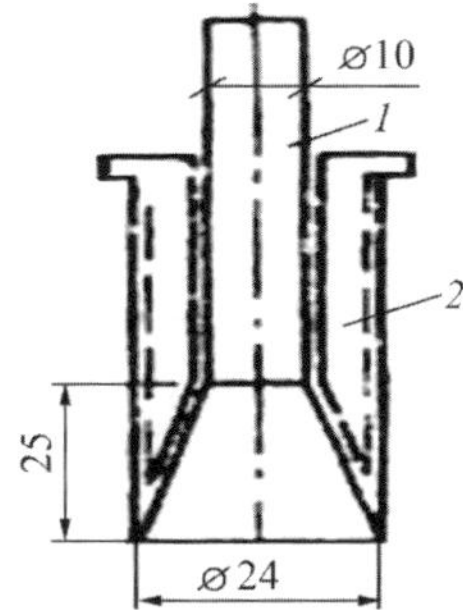

Fig. 7.8 The DYNA device
(Proceq SA, Switzerland) for
tearing testing anchors and
rods. [9]

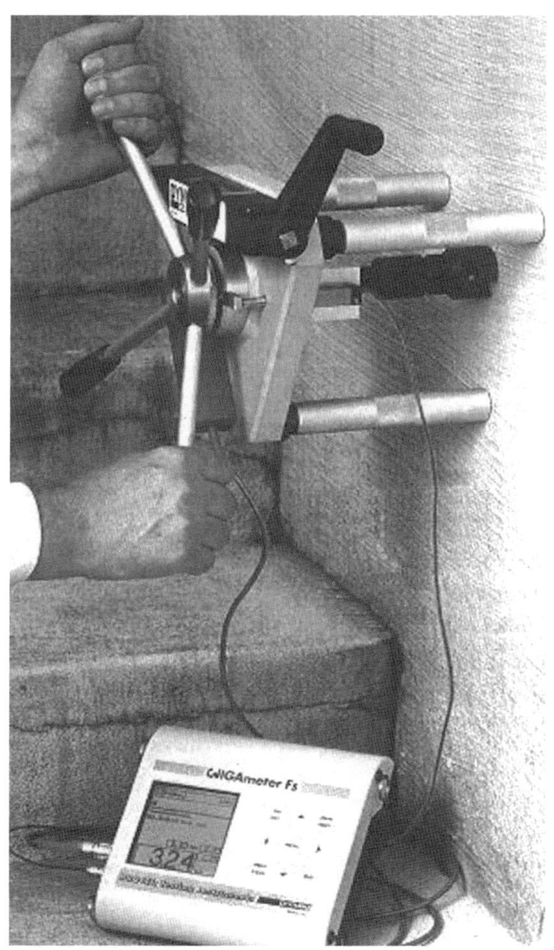

Similar devices are available from R&D Enterprise "Interpribor" (Chelyabinsk, Russian Federation) [10]. The purpose of the ONIKS-OS device (Fig. 7.9) is determination of concrete strength at cleavage tearing during the inspection of buildings and

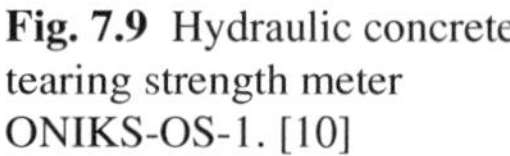

Fig. 7.9 Hydraulic concrete
tearing strength meter
ONIKS-OS-1. [10]

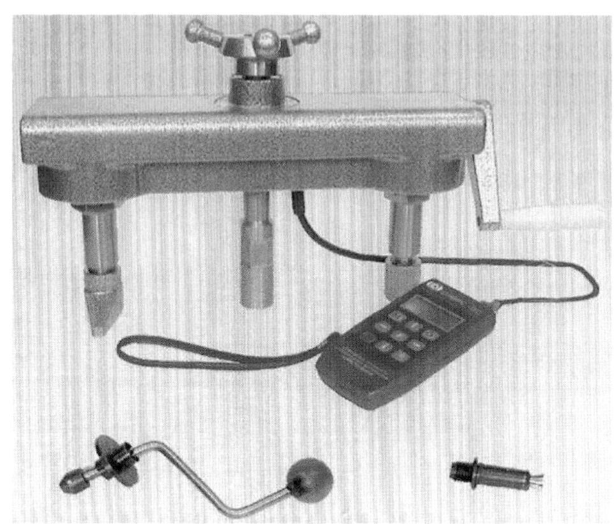

structures as well as rocks, beams, blocks, etc. It has the advantage of a two-cylinder
double-seat hydraulic press system with tearing axis self-adjusting along the anchor
fixture axis. Strength measurement range is 5.0 ... 100.0 MPa; maximum anchor
tearing force is 65 kN, and force measurement error is within 2 %.

The electronic module indicates force values, automatically fixes the tearing force,
calculates concrete strength, finds the mean value and coefficient of variation in a
series of five tests, saves results for 360 series with five measurements in each, and
provides data transfer to a personal computer [10].

Method of Structure Edge Chipping

Once more, a method of determination of concrete strength *in situ* requires chipping
of the concrete lump at the structure edge with fixation of (or recording) chipping
force in accordance with GOST 22690.4-77 [7]. The structure edge chipping depth
may be 10, 20 or 30 mm. *In situ* concrete strength R_b follows from a calibration
curve for chipping force $P_c \sim R_b$. In order to plot such a curve, a series of laboratory
specimens $20 \times 20 \times 20$ cm in size, made from different concrete grades, is subject
to testing (see Sect. 7.1). Each edge of every specimen undergoes only the single
chipping. Required concrete strength values are subject to establishment through
testing the laboratory specimens in a press.

Figure 7.10 shows an ONIKS-SR appliance specially designed for concrete
strength measurements using the method of edge chipping during the inspection
of buildings or structures [10].

This appliance has the virtue of an advanced fastening method using one or two
anchors 10 mm in diameter. In other words, it has no need to grip two angles of a
structure as with previously used models [8], opening the possibility of measuring

Fig. 7.10 ONIKS-SR appliances for concrete strength measurements using the method of edge chipping

in hard-to-reach spots. The hydraulically-driven, compact power device is equipped with a built-in electronic module for data processing.

Strength measurement range is $R_b = 5 \ldots 100$ MPa; maximum chipping force is $P_{max} = 35$ kN, and force measurement error is within 2 %.

The electronic module fixes measurement results, saves results for 360 series with five measurements in each, and provides data transfer to a personal computer.

7.2   Physical Methods and Devices for Technical Diagnostics of Long-term Concrete Structures

A great number of publications are devoted to the development and implementation of methods for diagnosing the integrity, defective status, and strength characteristics of concretes and similar building materials and structures based on the physical processes of deformation and fracture of materials [11–18]. Most general classification of types and physical concepts put into the base of methods for determining physical and mechanical characteristics of structural materials is presented in the reference book [2] (see table in the preface to the present chapter). The table, as well as the present section, describes only six of the most common methods of physical diagnostics of building materials and structures according to Ukrainian standard DSTU 2865-94 from thirteen given in [2]. These methods comprise those based on visual, optical, acoustic, thermography, radiation, and agent penetration concepts.

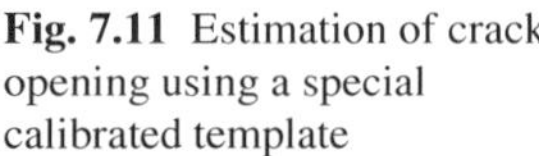

Fig. 7.11 Estimation of crack opening using a special calibrated template

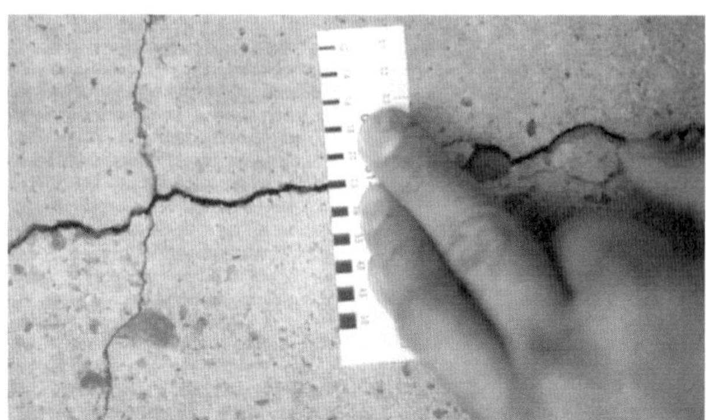

Visual Method

This method is applicable for revealing the relatively coarse defects, surface or through cracks, voids, etc. Visual inspection gives the primary information about the structure's status. This method is the most easily available, simple in conducting, and quite highly effective. Depending on visual acuity, illumination, and object contrast, the operator can reveal cracks with an opening as small as 0.1 mm (Fig. 7.11).

Optical Method

Optical instruments essentially enhance the possibilities of the human eye, allowing the observer to see otherwise invisible or irresolvable objects. Such instruments include magnifiers, microscopes, endoscopes, cathetometers, periscopes, and others.

Depending on the mode of light interaction with the observed structure, there are optical methods of observation in transmitted, reflected, diffused, or stimulated light. Informative parameters here are light amplitude, phase, degree of polarization, frequency spectrum, transmission time, angles of refraction and reflection, etc.

Instruments for Optical Method Implementation

Restricted access for inspection of some structural fragments often requires using instruments based on light guides. The operator places the optical probe into a bore drilled in the structural area under inspection. While moving the probe into the bore, the operator examines the bore walls and seeks defects (Fig. 7.12).

In order to inspect and reveal inner damages in various piping, video cameras mounted on self-propelled machines are used. State Engineer Center "Techno-Resurs" NASU in collaboration with Karpenko Physico-Mechanical Institute NASU developed the self-propelled diagnostic device consisting of video camera and electronic module for inspection of the damage status of concrete structures (Fig. 7.13).

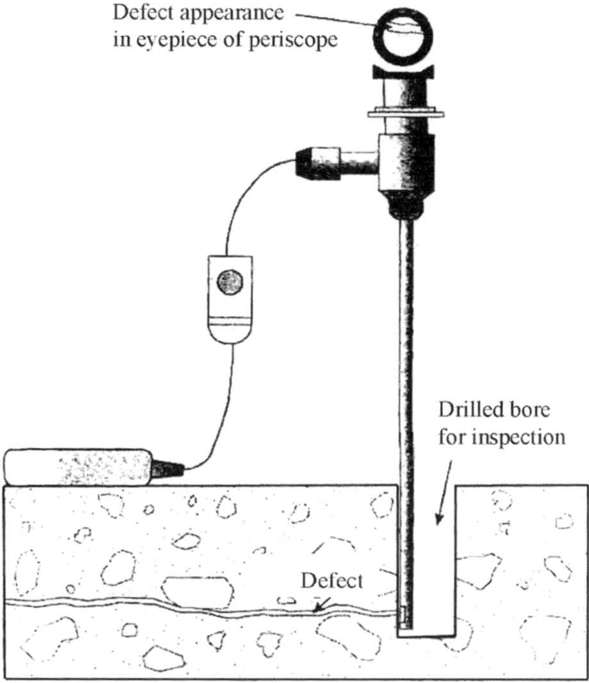

Fig. 7.12 Defect diagnostics in concrete walls using periscope. [4]

Fig. 7.13 Self-propelled video diagnostic system developed by SPRC "Techno-Resurs" NASU. [19]

Thermographic Method

The physical base of this method lies in the temperature gradients arising in a damaged object during its heating or cooling. Laminations, cracks, and cavities interrupt heat flows in the inspected structure. Therefore, temperatures of the

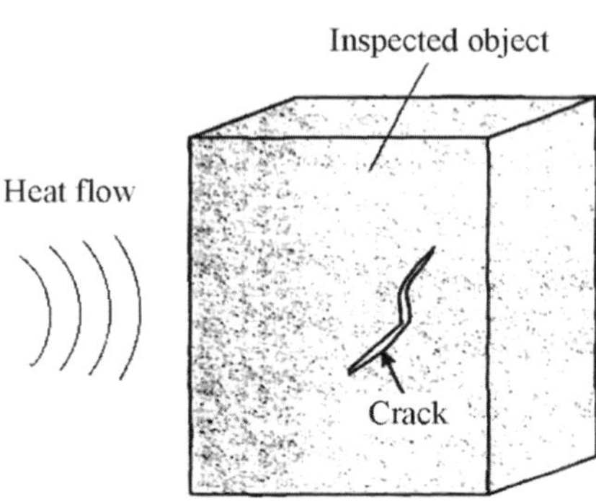

Fig. 7.14 Layout of defect control using heat flow distribution

concrete surface above the damage become higher during heating or lower during cooling, as compared with other surface areas.

There exist passive and active methods of thermography. In the passive method, any external heat sources are absent and only self-radiation of the structure is used. In the active method, the structure undergoes heating by an external heat source (Fig. 7.14).

In the practice of building structural diagnostics, the method of active thermographic control is more widely spread. A heat source imparts heat to the structure using a contact or contactless method. Temperature gradients on the surface are revealed using thermocouples, thermosensitive paints, or other methods. An infrared imager is one of the most effective means for temperature fields and heat flows visualization. Its physical base consists of the temperature dependence of heat radiation in the infrared and/or visible portion of the spectrum.

Radiation Method

The physical substantiation lies in the interaction of X-rays and gamma radiation (or, sometimes, other radiation types) with atoms of materials occurring in the way of its propagation. The more dense the material, the higher the intensity of radiation absorption. Consequently, penetrating radiation flux density increases in defect areas of the structure. Radiation methods can be radiographic or radioscopic, depending on data acquisition techniques [11]. The former forms optical images on X-ray films reproducing results of radiation interaction with the examined structure. This method is quite sensitive and appropriate for revealing defect sizes and positions.

The latter transforms the radiation pattern of the examined structure into a visible image in the display of a radiation-optical converter. This method is faster and less expensive than the former [11].

A novel and most promising method of radiation control, X-ray computed tomography (X-CT), has emerged in recent years. It combines the informative possibilities of X-ray radiation with up-to-date advances in computer and digital technologies.

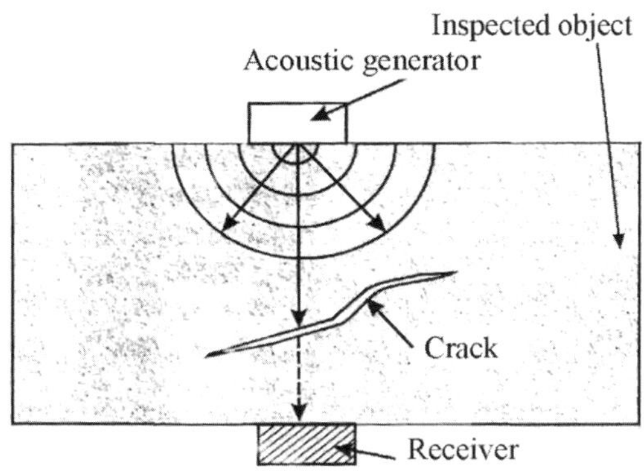

Fig. 7.15 Diagnostics of a crack-containing structure using ultrasonic wave transmission parameters

However, this method finds application only in the diagnostics of particularly important building structures due to the high cost of the respective equipment [12].

Acoustic Methods and Ultrasonic Inspection

Acoustic methods are based on elastic wave propagation in various media. For this purpose, vibrations with frequencies from 50 to 50 MHz are used. Accordingly, acoustic methods embrace vibrations of low frequency (sound and lower ultrasound ranges below 100 kHz) and high frequency (ultrasound from 100 to 50 MHz).

To study concretes, most specialists use the ultrasonic vibrations in the frequency range from 20 to 200 kHz. There is a close connection between propagation of elastic waves in a material and its physical and mechanical characteristics, such as elastic moduli, density, and so forth. The fact that the acoustic properties of air and solids are very different makes it possible to discriminate the ultrasound wave reflected from extremely thin defects such as cracks, poorly welded seams, etc.

Methods of acoustic inspection can be passive or active. The passive methods register vibrations arising in the structure in the course of operation (service) due to either intrinsic causes or extrinsic stimulation. The method of acoustic emission based on the phenomenon of elastic waves emission by a material during crack nucleation and/or growth [2], [15] may serve as an example of the passive acoustic methods.

The active acoustic methods are distinctive in that the acoustic vibrations in the structure come from outside. Parameters of the induced acoustic field are subject to recording and analyzing after interaction with the structure's area of interest. The recording and analyzing techniques for penetrating, reflected, and/or combined radiation are different.

In particular, techniques of penetrating radiation (Fig. 7.15) include:

a. Amplitude-shadow technique based on registration of the ultrasonic wave amplitude change after transmission through the inspected structural area;

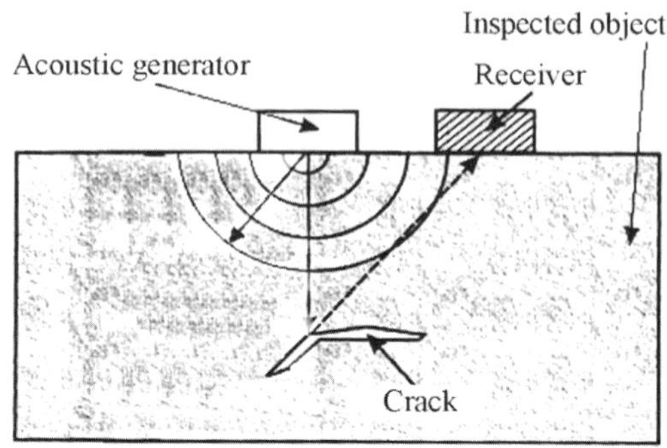

Fig. 7.16 Diagnostics of a crack-containing structure using reflected wave technique

b. Time-shadow technique intending registration of the wave transmission time increment due to rounding of the defect, and
c. Velocity metric technique providing registration of wave propagation velocity changes in the defective material.

Techniques of penetrating radiation are applicable when the structure is accessible from both sides. These techniques reveal cracks located parallel or nearly parallel to the base surface.

In the absence of two-side access to the structure, the techniques of reflected radiation (echo method) are preferable (see Fig. 7.16). They allow determining defect size and position in a concrete mass as well.

Combined techniques imply registration and analysis of both reflected and penetrating waves. (For more details, see references in [2], [17].)

Devices for Ultrasonic Diagnostics of Concretes [10]

According to GOST 17621-78, the ultrasonic method is applicable for determining the strength of heavy concretes and concretes with porous aggregates, as well as cellular and dense silicate concretes. The method consists of measuring the ultrasound transmission velocity (v) in the material related to its elastic modulus and density (ρ) by the following formulas [7]:

$$v = \left(\frac{E_g}{\rho} \right)^{1/2} - \text{for longitudinalwaves;}$$

$$v = \left(\frac{G_g}{\rho} \right)^{1/2} - \text{for transversewaves;}$$

where E_g and G_g are the dynamic Young and shear moduli defined as the stress-strain curve tangent slope near the origin (see Sect. 7.1) and ρ is the concrete density.

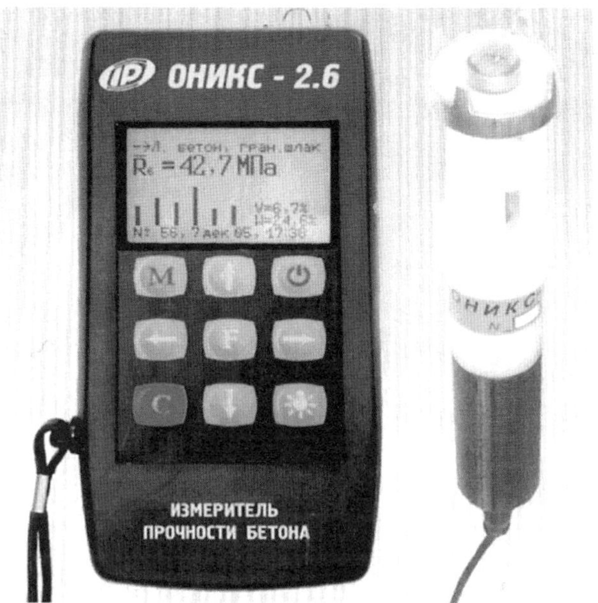

Fig. 7.17 Concrete strength meter and defectoscope ONIKS-2.6. [10]

The concrete strength is subject to determination using measured values of acoustic wave (sound) velocity and an independently measured calibration curve $R_b(v)$.

The advanced ultrasonic devices have the calibration curves $R_b(v)$ saved in memory together with software for identification of defect size and position as well as calculation of the concrete's compressive strength.

Defectoscope ONIKS-2.6 with Strength Measurement

This device performs the object defectoscopy by comparing signals from the structure and reference specimens in response to a shock wave or its components. The device ONIKS-2.6 enables us to take into consideration the concrete's age, composition, and hardening and carbonation mode. The device has 60 basic calibration curves for various concrete grades saved in memory [10].

The strength measurement range is $R_b = 0.5 \ldots 100$ MPa with a permissible error of 7 %. The shock energy of the primary acoustic generator is within $0.1 \ldots 0.12$ J.

The device can save in memory roughly 31 thousand results and processes.

As a rule, advanced ultrasound defectoscopes have the design of all-purpose instruments. For example, the ultrasound defectoscope Pulsar-1.2 (Fig. 7.17) enables us to reveal any voids, cracks and/or other defects nucleated in the stages of manufacture and use.

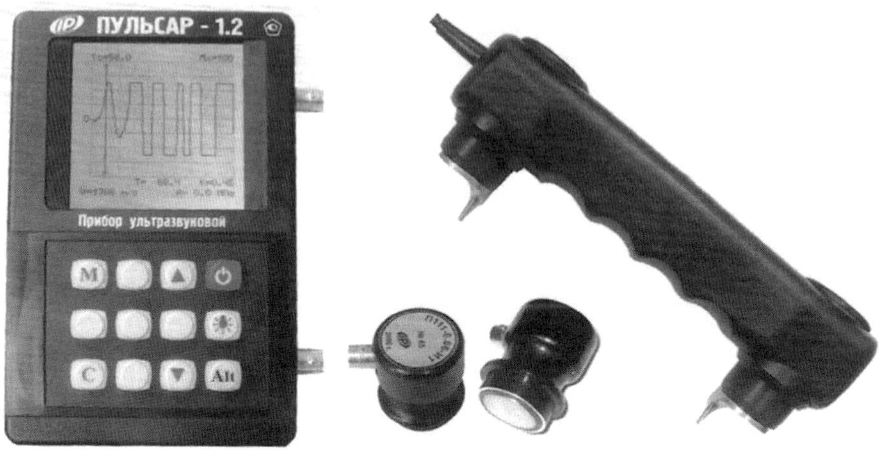

Fig. 7.18 Ultrasound defectoscope Pulsar-1.2 with surface acoustic generator. [10]

At the same time, Pulsar-1.2 provides monitoring of strength, Young modulus, density, and degree of concrete inhomogeneity, as well as measurement of crack depth in concrete and reinforced concrete products, and certain other functions (Fig. 7.18).

Proceq SA (Switzerland) [20] produced similar ultrasonic devices in the TICO series (Fig. 7.19) that is, at present, substituted with the advanced series Pundit® Lab and Pundit® Lab Plus. These devices enable us to reveal voids, cracks and/or other defects that arose during operation as well as determine concrete strength, Young modulus, and density. The devices represent the obtained information in an LCD display, save data in their memory (up to 500 measurements) and allow for transfer to a personal computer.

Methods of Impregnation (Agent Penetration)

This class of diagnostic methods includes two basic groups: capillary and leak detective. The former uses the capillary infiltration of indicative liquids into the porosity systems of the surface defects. The defects become visible as optically contrasting patterns on the structural surface after filling the free volume with the indicative liquids (Fig. 7.20). The visibility is made possible by the fact that the width of the so-formed indicative lines is much higher than the opening of the defects.

Paraffin oils, such as kerosene, often play the role of the penetrating agent. Kerosene easily penetrates into pores and cracks due to small viscosity and high wettability. Agents able to luminesce under ultraviolet radiation find a wide application as well.

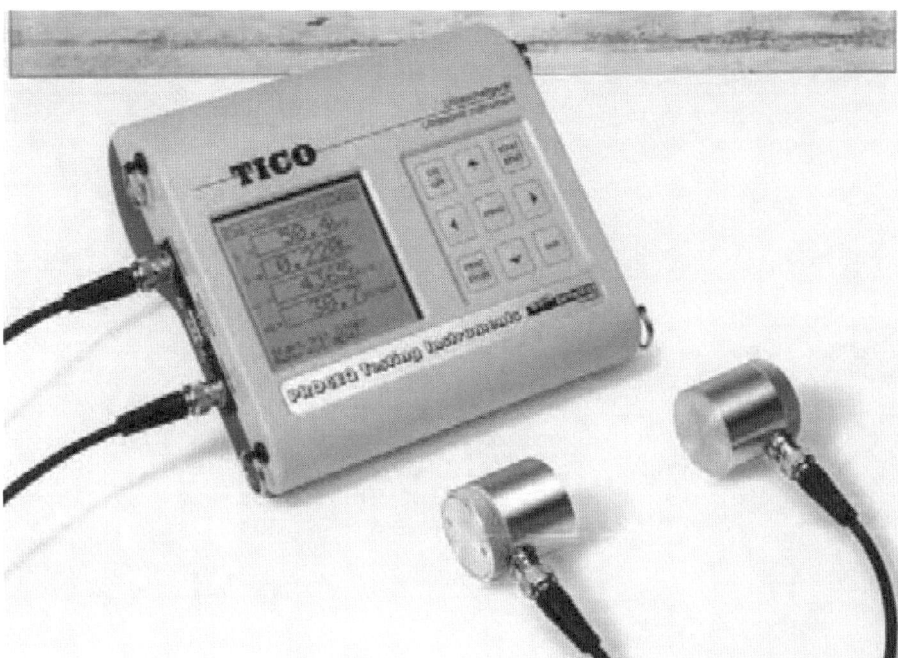

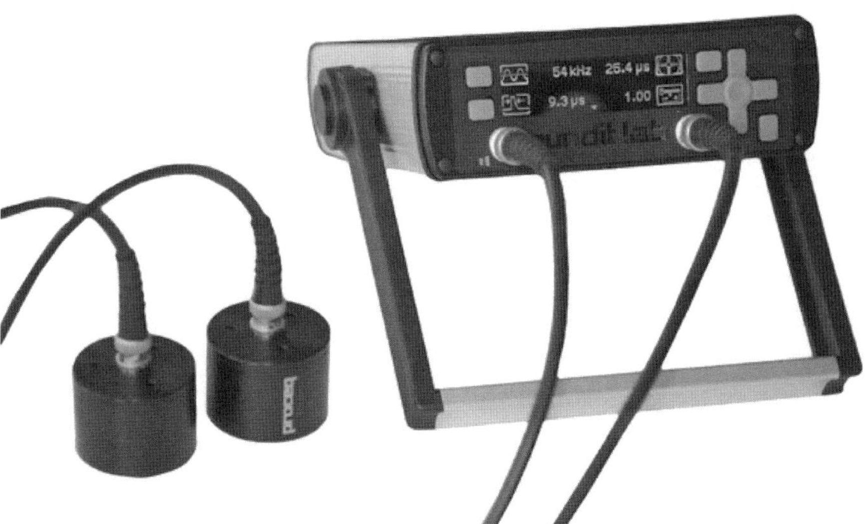

Fig. 7.19 Ultrasound defectoscopes of Proceq SA (Switzerland) [20]: TICO (*top*) and Pundit® Lab (*bottom*)

The latter method of leak detection is useful for revealing the thorough defects. Low viscosity liquids or gases here play the role of penetrating agents. One modification consists of blowing the structural surface with compressed air or an air/nitrogen

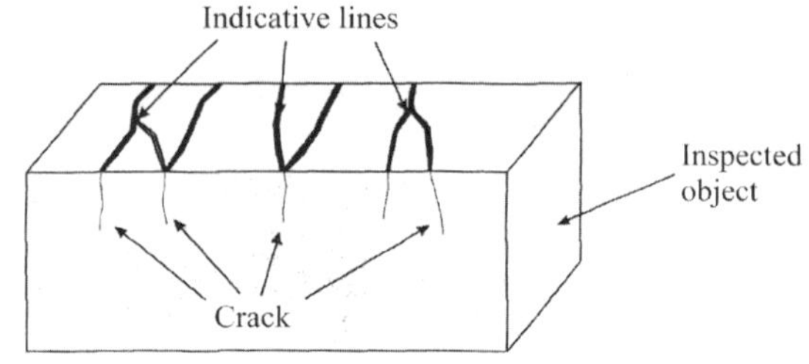

Fig. 7.20 Defect diagnostics using penetrating agents

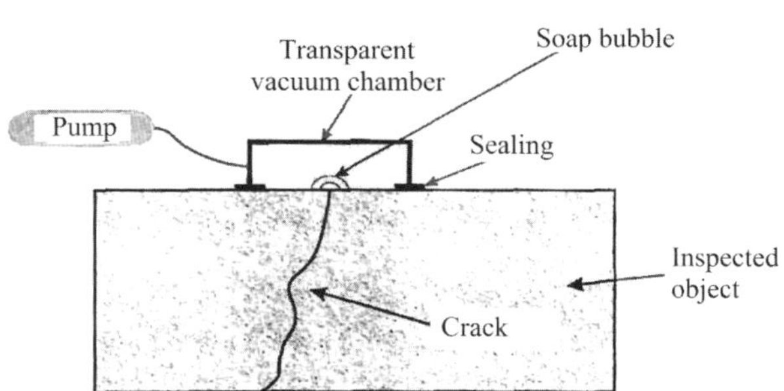

Fig. 7.21 Structural element diagnostic using the vacuum method

mixture possessing especially high penetrability. The opposite side of the structure is subject to wetting by soapsuds. Bubbling indicates the presence of through defects. Another modification includes blowing with ammonia. Instead of soapsuds, the phenolphthalein covering the opposite side of the object serves as the indicator in this case. A crack becomes visible due to a dark spot on the opposite surface covered by the indicator. Sometimes ultrasound receivers can supersede the indicator by registering vibrations arising due to air escaping from cracks. Finally, one more modification involves revealing through cracks using a vacuum (Fig. 7.21).

A transparent vacuum chamber is subject to attachment to the structural surface wetted by a soap solution. A soap bubble appearing during air pumping off the chamber indicates the presence of a crack. Besides kerosene, some other high fluent liquids are applicable for revealing through cracks, for example, turpentine [15].

7.3   Electrochemical Tools and Methods of Chemical Diagnostics

Electrochemical diagnostic methods are comprised of electric resistance measurements, pH measurements, electric potential mapping, and impedance studies of concrete structures (see table in the preface to the present chapter).

Fig. 7.22 Concrete resistivity meter RESI (lately Resipod) with 4-point Wenner probe of Proceq SA (Switzerland). [21]

Electric Resistance

Measuring the electric resistivity of reinforced concrete structural elements provides information on the intensity of the reinforcement surface corrosion. Corrosion products (rust) migrate into the bulk of the concrete. Ferrous oxides and/or hydroxides have relatively high electric conductivity and therefore reduce the resistivity of adjoining sections of concrete. As experimentally established, reinforcement corrosion is virtually improbable at concrete resistivity $\rho \geq 12$ kOhm cm. However, reinforcement corrosion is probable at resistivity $\rho = 8 \ldots 12$ kOhm cm and very probable at $\rho < 8$ kOhm cm.

To measure the electric resistivity of concrete in local areas, the ohmmeters operating in the kiloohm range equipped with special electrodes (Fig. 7.22) are applicable [21].

Electrodes for concrete resistivity measurement consist of platinum rods connected to inert conductors (graphite, coal) placed into a certain conductive medium (e.g., an aqueous solution of 3 % NaCl). A special material capable of accumulating moisture and supplying it to the concrete surface retains the medium near the conductive rods. SPRC "Techno-Resurs" NASU have developed and designed one modification of such electrodes [22].

In particular, the concrete resistivity meter RESI by Proceq SA (Switzerland) has the measured resistivity range $\rho = 0 \ldots 99$ kOhm cm and measurement error $\Delta\rho = \pm 1$ kOhm cm. The RESI indicator module has a memory up to 7,200 measured values, LCD display 128×128 dpi and original software for data processing and transfer to a personal computer.

Method of Electric Potential Measurement

Combined measurements of electric resistance and electric potential distribution yield more detailed information about reinforcement corrosion. Steel corrosion in concrete is an electrochemical process and, hence, induces the galvanic effect. For such processes, the special electrodes, microgalvanic couples, exist, enabling us to

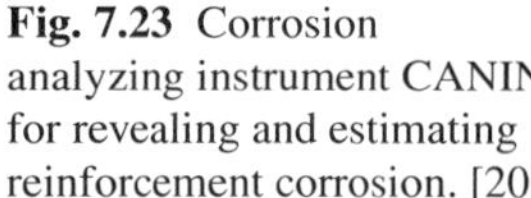

Fig. 7.23 Corrosion analyzing instrument CANIN for revealing and estimating reinforcement corrosion. [20]

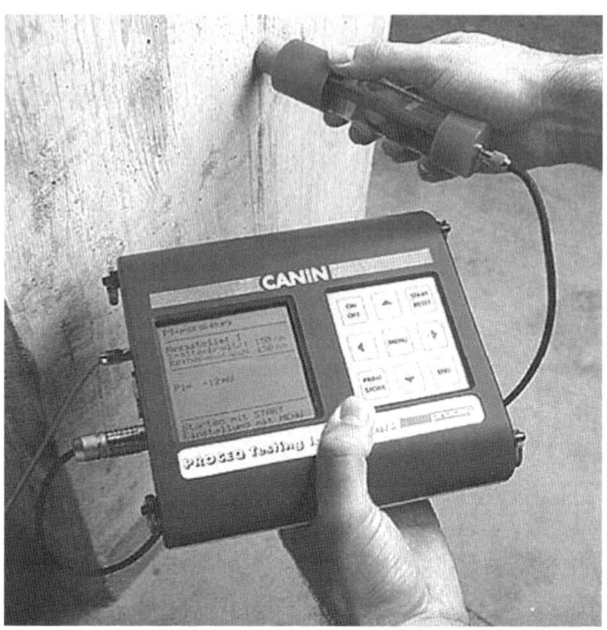

measure currents arising inside the structure. Mapping such currents across a whole surface, one can find the reinforcement corrosion areas.

A number of worldwide firms produce different instruments for non-destructive survey of reinforcement corrosion in building structures, as well as revealing rust at the early stages when it rarely manifests visual signs and/or causes concrete destruction. Particularly, Proceq SA (Switzerland) proposes devices of series CANIN [20] (Fig. 7.23) permitting estimations of reinforcement status across large surface

Fig. 7.24 Potential mapping across floor and overhead surfaces. [20]

areas by means of direct data representation in built-in display. Electrode wheels are applicable for horizontal, vertical, inclined, or overhead surfaces.

Figure 7.24 illustrates the instrument application for diagnosing reinforcement corrosion in building structures where 4-point electrode wheels build maps of floor and overhead concrete elements.

The number of electrode wheels in a system can vary between 1 and 8. Each electrode wheel has a subsystem of permanent wetting for the concrete surface along the total path up to 200 m.

The LCD display has a resolution of 128 × 128 pixels and backlight. Memory space is high enough to store up to 5,800 individual measurements (CANIN+). One-time corrosion analysis is available for a concrete area over $100\,m^2$ without connecting to the computer.

References

1. Paton BE (1997) Non-destructive control and reliability of technical solutions. Bull Acad Sci Ukr SSR 1:71–76
2. Nazarchuk ZT (ed) (2001) Neruinivnyi kontrol' i mitsnist' materialiv (Non-destructive testing and strength of materials). In: Panasyuk VV (ed) Mekhanika ruinuvannya materialiv i mitsnist' konstruktsii (Fracture mechanics of materials and strength of structures). vol 5. Karpenko Phys. Mech. Inst. NASU, Lviv
3. Panasyuk VV (ed) (1988) Mekhanika razrusheniya i prochnost' materialov (Fracture mechanics and strength of materials). In: Panasyuk VV (ed) Mekhanika ruinuvannya materialiv i mitsnist' konstruktsii (Fracture mechanics of materials and strength of structures). Nauk. Kyiv, dumka
4. Czarnecki L, Emmons PH (2002) Naprava ∇ ochrona konstrukcji betonowych (Repair and protection of concrete structures). Polski Cement, Krakov
5. Marukha VI (2007) Strength of concretes with stress concentrators filled by polyurethane mixtures. In: Mekhanika i fizika ruinuvannya budivel'nykh materialiv i konstruktsii (Fracture mechanics and physics of building materials and structures). Issue 8, Kamenyar, Lviv, pp 223–232
6. Standard of Ukraine DSTU 2865-94 (1995) Kontrol' neruinivnyi (Non-destructive testing). Derzhstandart Ukrainy, Kyiv
7. Leshchinskii MY (1980) Ispytaniya betona (The concrete testing). Stroyizdat, Moscow
8. Leshchinskii MY, Chermenyan MP, Khutoryanskii MS et al (1980) Spravochnik rabotnika stroitel'noi laboratorii zavoda ZhBI (Handbook of researcher in reinforced concrete works laboratory). Budivel'nik, Kyiv
9. Pull Off Tester—Dyna. Proceq SA (Switzerland). http://www.procEq.com/en/non-destructive-test-equipment/concrete-testing/pull-off-testing/dyna.html. Accessed 13 Apr 2013
10. Concrete hardness tester(separation)ONIKS-OS. OOO NPP Interpribor (Chelyabinsk, Russia). http://interpribor.com/?id=7721. Accessed 13 Apr 2013
11. Radiofizicheskiye metody kontrolya v stroitel'stve (1973) (Radiophysical monitoring methods in construction). Stroyizdat, Leningrad
12. Nerazrushayushchiye metody ispytanii stroitel'nykh materialov i konstruktsii (1982) (Non-destructive testing methods for building materials and structures). Riga Polytech Inst, Riga
13. Sovremennyye fizicheskiye metody i tekhnicheskiye sredstva kontrolya kachestva stroitel'nykh materialov i konstruktsii (1983) (Modern physical tools and methods for building materials and structures quality control). Rus Soc "Znaniye", Moscow
14. Albrecht R (1979) Defekty i povrezhdeniya stroitel'nykh konstruktsii (Defects and damages in building structures). Stroyizdat, Moscow

15. Lushin OV, Volokhov AV, Shmakov GB et al (1985) Nerazrushayushchiye metody ispytanii betona (Non-destructive testing of concrete). Stroyizdat, Moscow
16. Sammal OY, Gul'kov AA, Mihkelson RA (1981) Sklerometricheskii mikroprotsessornyi izmeritel' VSM-4 dlya opredeleniya prochnosti betona (Sclerometric microprocessor meter VSM-4 for measurements of concrete strength). R & D Constr Est SSR, Tallinn
17. Ahverdov IN, Margulis LN (1975) Nerazrushayushchii kontrol kachestva betona po elektroprovodnosti (Non-destructive electric conductivity quality testing of concrete). Nauka i Tekhnika, Minsk
18. Problemy naukowo-badawcze konstrukcji z betonu (1999) Research problems of concrete structures. Polski Cement, Krakov. P. 232
19. Marukha VI, Serednitskiy YA, Gnip IP, Sylovanyuk VP (2007) Development of injection technologies and design of mobile equipment for diagnostics and serviceability restoration of concrete or reinforced concrete structures operating under stress-corrosive conditions. Sci Innov 3:55–62
20. Ultrasonic Pulse Velocity—Pundit Lab Plus. Proceq SA (Switzerland). http://www.procEq.com/en/non-destructive-test-equipment/concrete-testing/ultrasonic-pulse-velocity/pundit-lab-plus.html. Accessed 13 Apr 2013
21. Resipod Resistivity Meter. Proceq SA (Switzerland). http://www.procEq.com/en/non-destructive-test-equipment/concrete-testing/moisture-corrosion-analysis/resipod.html. Accessed 13 Apr 2013
22. Gnyp IP, Neprila MV, Mertsalo IP (2006) Electrode for measuring electrochemical characteristics of protective coverings. Declarative patent of Ukraine UA12978, 15 Mar 2006

Chapter 8
Implementation of Injection Technologies in the Renewal and Restoration of Serviceability of Concrete and Reinforced Concrete Structures

Abstract This chapter presents results of wide practical implementation of injection materials and technologies for renewal of concrete and reinforced concrete structures in industrial and municipal building structures of Ukraine damaged by cracks and/or other defects. Case studies of combined injection and waterproofing works in overground, underground, and underwater structures of stationary atomic power plants, as well as river hydroelectric pumped storage power stations, are illustrated. Descriptions of blockage technologies of water leakage through cracks or seepage through defect zones using injection materials in dams, turbine rooms, or adjoining working rooms of hydraulic works are represented. The process specific features and practical results on renewal of inner surfaces in concrete and reinforced concrete sewage collectors and water main pipelines 1400...3000 mm in diameter in the city of Lviv are given. Case studies of injection strengthening and surface protective waterproofing of injured tunnels, bridges, and traffic interchanges for rail and motor transport, as well as foundations, walls, and vaults of houses, churches, and other historical relics in antique city districts, are presented. Renewal with modern cement materials and surface protective waterproofing of reinforced concrete berths and other structures in seaports that have undergone long-term exposure to sea slaps and active humid air corrosion is illustrated.

Many researchers and engineers around the world have applied great effort to the development and implementation of new injection materials and technologies for the renewal and restoration of serviceability of concrete and reinforced concrete structures [1–8].

Over a period between 2002 and 2008, the specialists of the State Engineer Center "Techno-Resurs" NASU in collaboration with the Karpenko Physico-Mechanical Institute NASU developed and implemented novel injection compositions and technologies in Ukraine. Among the most prominent achievements were the hydroelectric pumped storage plants (HPSP) Tashlytska and Novodnistrovska of the National Energy Company "Energoatom", the Lviv municipal water duct, and the Illichevsk seaport. The implemented technologies mainly included injection of novel polyurethane compositions and auxiliary treatment of the renewed areas with finishing waterproof materials. Several cases also included repairing surface damages in concrete and reinforced concrete structures using cement-polymer mortars [2],

V. V. Panasyuk et al., *Injection Technologies for the Repair of Damaged Concrete Structures,* DOI 10.1007/978-94-007-7908-2_8,
© Springer Science+Business Media Dordrecht 2014

Fig. 8.1 General appearance of the Tashlytska HPSP in Pivdennoukrainsk, Mykolayiv region of Ukraine

[4–6]. The abundant technical information concerning applications of modern materials and technologies for the injection strengthening as well as waterproofing of concrete, reinforced concrete, and brick structures and buildings in various branches of the economy exists in special literature and datasheets from leading European firms [9–12]. This section presents specific case studies of the implementation of injection materials and technologies.

8.1 Injection Renewal of Damaged Concrete and Reinforced Concrete Structures of Hydraulic Constructions, Cooling Towers, and Atomic Power Plants

Tashlytska HPSP

Two particularly successful applications of injection materials and technologies took place at the hydraulic structures of the Tashlytska HPSP, NEC "Energoatom of Ukraine", in Pivdennoukrainsk, in the Mykolayiv region of Ukraine (Fig. 8.1, see Appendix) and the Novodnistrovska HPSP, JSC Ukrgydroenergo, in the Chernivtsy region, Ukraine. One specific issue that arose in the Tashlytska HPSP was water leakage through cracks and damages in joints between concrete blocks of the dam (Fig. 8.2). These joints underwent combined strengthening, waterproofing, and

Fig. 8.2 Areas of water leakage through cracks and damages in joints between concrete blocks of the Tashlytska HPSP dam

Fig. 8.3 Area of intense wetting on the reinforced concrete dam wall with inserted packers for feeding the injection polyurethane compositions

serviceability renewal through the introduction of pressurized (50 . . . 150 atm) fluent polyurethane compositions into the cracks, discontinuities, and damages of the reinforced concrete dam in its lower underwater section 16 m in height [5], [6].

Water leakage and seepage through the reinforced structures of the dam was stopped (see Figs. 8.2 and 8.3) over the total area of 10 000 m^2. Figure 8.4 in the Appendix illustrates the procedure of injecting fluent polyurethane compositions

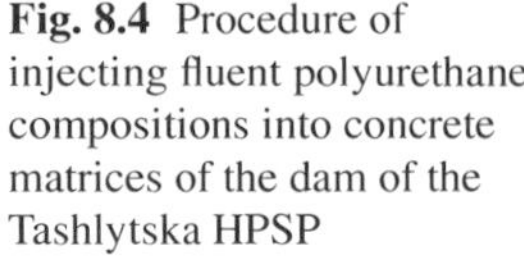

Fig. 8.4 Procedure of injecting fluent polyurethane compositions into concrete matrices of the dam of the Tashlytska HPSP

under pressure of 50 ... 150 atm into the subsurface cracks and damages through the guiding bores (packers) pre-drilled into the concrete.

A specificity of the work execution under strong water leakage or even weak water seepage through cracks and/or defects in the concrete consists of necessary water stoppage in cracks and damages in the reinforced concrete blocks of the dam. Therefore, these defects were initially subject to filling with the foaming polyurethane compositions through drilled bores. During foaming, the fluent polyurethane materials expanded in volume by 10–30 times [6], [8], [11].

Reacting with the water and foaming, the injected polyurethane compositions quickly structurized (over 20 ... 60 s) by forming solid water-resistant polyurethane foam. In such a way, the tight filling of the entire free volume in cracks and defects as well as high polyurethane foam adhesion to the inner concrete surfaces enabled the complete closure of water leakage and seepage paths. Subsequently, 10 ... 20 min after the first injection, the polyurethane foam inserts in the cracks and damages were subject to strengthening by feeding the fluent non-foaming polyurethane compositions into the same packers [5], [12].

Strength and Serviceability Renewal in Reinforced Concrete Shafts of Turbine Rooms

The issue that arose at the Novodnistrovska HPSP consisted of serviceability restoration of a shaft of reinforced concrete blocks in the connection that became impaired because of protracted construction. The impairment had resulted in water penetration to the reinforced concrete shafts of the turbine rooms (Fig. 8.5). Figure 8.6 shows the general scheme of the arrangement and connection of reinforced concrete hydraulic structures in the HPSP.

Fig. 8.5 The reinforced concrete shaft of the turbine room in the Novodnistrovska HPSP subjected to injection strengthening and blocking of water seepage areas

The structure had undergone a complete set of repair and renewal procedures ensuring elimination of subsurface cracks and damages in the reinforced concrete, as well as blockage of water seepage and leakage areas in the concrete matrix. For this purpose, the same fluent polyurethane injection materials (foaming *and* non-foaming), process technologies and equipment were used as those applied in the reinforced concrete storage dam and adjoining reinforced concrete working rooms of the Tashlytska HPSP [2], [5], [6].

Similar renewal procedures for water blockage in cracks and damages of concrete matrices find wide application in foreign countries as well. For instance, specialists of DESOI GmbH (Germany) stopped water leakage and sealed the power dam in the river canyon near Neustadt (Germany), see Figs. 8.7 and 8.8 in Appendix [6], [11], [12].

1. Water intake
2. Vertical underwater conduits
3. Horizontal conduits
4. HPSP room
5. Discharge conduits
6. Water outlet
7. Upper drainage gallery
8. Lower drainage gallery
9. Drift # 1
10. Drift # 2
11. Drift # 3

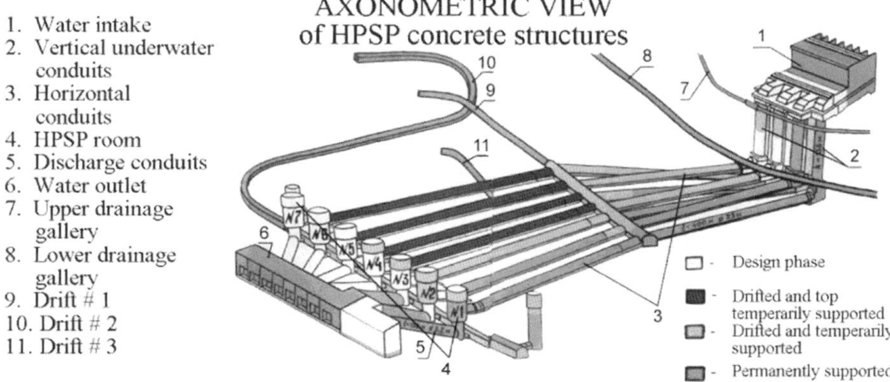

Fig. 8.6 Axonometric view of the main hydraulic structures in the Novodnistrovska HPSP

Fig. 8.7 Water stream leakage through damage in stone dam near Neustadt (Germany) that was using the injection method

Fig. 8.8 General appearance of power dam near Neustadt. (Germany, photo courtesy of the company DESOI GmbH)

Cooling Towers

Case studies of implementing injection technologies for strengthening and serviceability renewal of major structures of thermal or atomic power plants such as cooling towers present in both domestic and foreign practice. In 2005, the specialists of the State Engineer Center "Techno-Resurs" NASU accomplished repair and renewal works in reinforced concrete structures of the cooling tower belonging to the closed corporation Lukor (Kalish, Ivano-Frankivsk region, Ukraine). The procedures included additional covering of the tower's inner surface with a protective heat-resistant cement-polymer layer besides injecting the damaged areas with polyurethane [5], [13]. The companies DESOI GmbH (Germany) and JSC KMK-Energo (Novokuznetsk, Russia) renewed similar structures, the cooling tower of a power plant and the reinforced concrete stack of a heating plant, respectively.

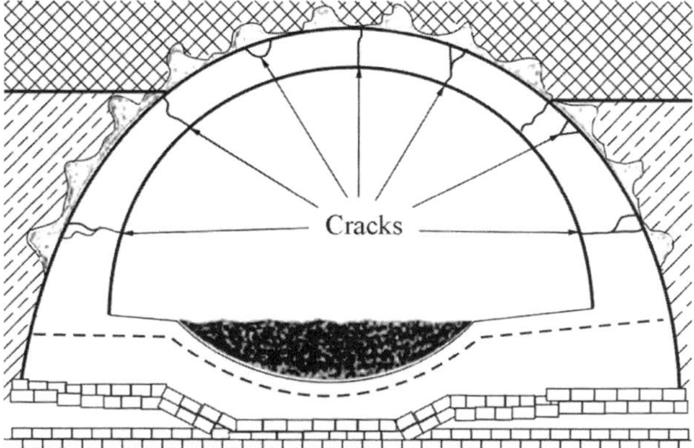

Fig. 8.9 General scheme of stress corrosion cracking in Lviv's concrete sewage collector at Svobody Avenue

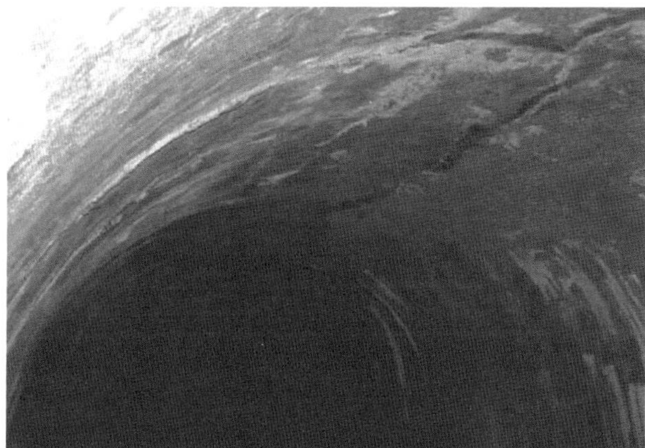

Fig. 8.10 Stress corrosion cracking (fracturing) in sewage collector under Svobody Avenue. (Lviv, Ukraine)

Renewal of Large-diameter Concrete Sewage Collectors

As the practice of serviceability restoration in municipal concrete and reinforced concrete sewage collectors of the public utility company Lvivvodokanal have shown, injection technologies are a most effective means in such restoration. Figure 8.9 shows the general view of the underground sewage collector in the city of Lviv built at the beginning of the 1900s. Lengthy service life had caused impairment of the concrete constituent of the structure's cupola roof. Namely, cracks and other injuries appeared in the concrete matrix (Fig. 8.10 of the Appendix). The stress corrosion cracking and fracturing processes over 100 years of use resulted in caving and failures in several sections of the municipal waste disposal system [5], [6].

To restore serviceability of the structure, composite fillers were required capable of ensuring appropriate adhesion to the system of concrete-polyurethane-concrete.

The specific feature of injection damage healing in the given case again consisted of the necessary interaction of fluent polyurethane materials with predominately humid and sometimes water-covered concrete surfaces. The technical approach was

Fig. 8.11 Drilling guiding bores in zone of crack in concrete vault

Fig. 8.12 Roof of a sewage collector repaired using injection technology and waterproofing

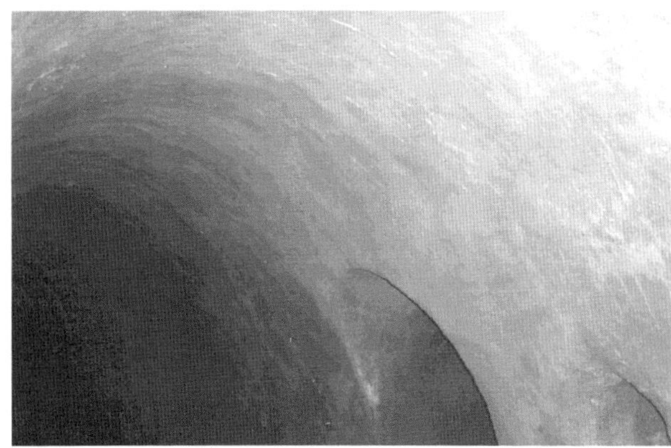

in blocking the water ingress by injecting the foaming polyurethane composition. Thereafter, non-foaming injection materials in new proportions had hardened the polyurethane foam in the cracks and damages [2], [4–6] (Fig. 8.11 in the Appendix). The executed renewal and repair procedures provided further reliable service of the sewage collector (Fig. 8.12). The same renewal and repair procedures using injection technologies subsequently provided renewal of collectors in the streets of Ivana Franka, Mitskevicha, and Bogdana Khmel'nitskogo of the city of Lviv. It should be noted here that cut-and-cover repair of street sewage collectors in the city core encountered significant difficulty due to the inevitable hindrance of urban traffic. In this view, the proposed injection technologies are indispensable in similar cases.

The experience gained by the specialists of the State Engineer Center "Techno-Resurs" NASU had shown that prolonged service life of concrete structures renewed by injection of fluent polyurethane compositions is possible only after applying 2–3 layers of waterproofing cement-polymer mortars onto the repaired surfaces. A good performance by the waterproof coating material AQUAFIN®-2K made by Schomburg GmbH & Co. KG (Germany) had demonstrated it to be especially suitable for

Fig. 8.13 Polymer injection waterproofing of reinforced large-diameter concrete water main in the reinforced concrete case. (Netherlands, photo courtesy of the company MC-Bauchemie Müller GmbH & Co. KG)

underground structures transporting water containing chemically active substances [5], [13].

However, waterproofing with modern cement-polymer mortars is also preferable in other structures [13]. In particular, it is applicable in repairing berths and other concrete or reinforced concrete structures in seaports (see below, the Illichevsk seaport).

Figures 8.13 and 8.14 in the Appendix illustrate examples of restorative works fulfilled by the German companies MC-Bauchemie Müller GmbH & Co. KG and DESOI GmbH using injection renewal and waterproofing of concrete and reinforced concrete collectors, water mains, and traffic tunnels. The foreign firms had used similar restorative materials and technologies in injection and waterproofing procedures.

8.2   Injection Renewal of Non-hydraulic Constructions, Tunnels, Bridges, and Historical Relics

Traffic Tunnels

Underground reinforced concrete tunnels operate under specific conditions, including static loading by ground mass above the structure and vibration loads during use

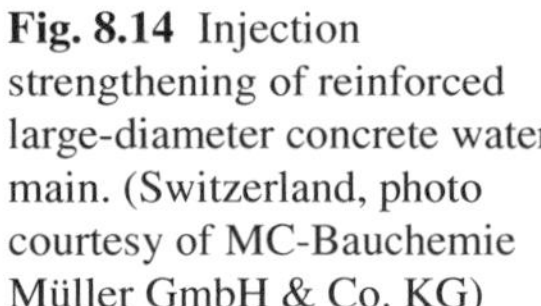

Fig. 8.14 Injection strengthening of reinforced large-diameter concrete water main. (Switzerland, photo courtesy of MC-Bauchemie Müller GmbH & Co. KG)

by vehicles; corrosive exhaust gases inside the tunnel and chemically reactive media outside it; temperature variations, etc. As a result, the concrete matrices of such structures tend to crack (Fig. 2.7 in Appendix). Similar conditions are characteristic for tunnels of underground railways. Therefore, the repair-and-renewal operations are required to prolong the service life of such structures (Fig. 8.15 in Appendix) [5], [6].

These operations often require injection materials and technologies. Thus, Figs. 8.16 and 8.17 of the Appendix illustrate the injection strengthening of reinforced concrete in the underground railways of Munich (Germany) and Kiev (Ukraine). Injection of fluent polymer materials into cracks and damages in concrete or reinforced concrete structures provides renewal and sealing of house footing and substructures, for example, foundations of car garages (Fig. 8.18 in Appendix).

Conservation of Historical Relics

Restoration works connected with the strengthening and renewal of historical relics and architectural landmarks have an obligatory aspect of preservation of the original external appearance. Injection technologies combined with minimum waterproofing

Fig. 8.15 Injection strengthening of damaged sections in Western Scheldt tunnel. (Germany, photo courtesy of DESOI GmbH)

Fig. 8.16 Injecting a fluent polymer composition into reinforced concrete wall of underground railway in Munich. (Germany, photo courtesy of DESOI GmbH)

Fig. 8.17 Drilling guiding bores for injection of fluent polyurethane compositions into reinforced concrete wall of underground railway in Kyiv, Ukraine

Fig. 8.18 Results of injection strengthening and waterproofing of the floor of an underground car park

Fig. 8.19 Repair-and-renewal operations in the foundation of the Museum of Ukrainian Arts on Dragomanova Street in Lviv

Fig. 8.20 Injection of fluent polyurethane compositions into the brick foundation of the Museum of Ukrainian Arts on Dragomanova Street in Lviv

of brick joints are one way of meeting the requirement of the smallest possible substitution of ancient building materials (stone, bricks, plasters, etc.). Figures 8.19 and 8.20 in the Appendix present a case of such repair-and-renewal operations based on injection technologies for filling dry cracks and damages in the subsurface of

Fig. 8.21 Waterproofing and additional protection of the foundations of the museum after polyurethane composition injection using modern cement-polymer materials

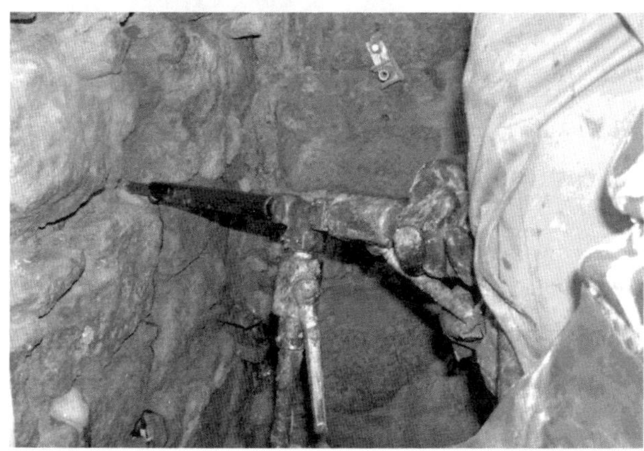

Fig. 8.22 Injection of polyurethane compositions into stress corrosion cracks (left and down from injection nozzle) lying deep in the stone masonry of the hotel "Leopolis" basements

brick and stone masonry as well as water seepage blockage in the inner surface of the foundation of the Museum of Ukrainian Arts in Lviv on Dragomanova street, 42. The SPC Center "Techno-Resurs" NASU was the executor of these restorative works as well (Fig. 8.21).

The materials and equipment used for restoration of historical relics were the same standard fluent polyurethane compositions and appliances used previously in the reinforced concrete dams and adjoining rooms of the Tashlytska and Novodnistrovska HPSPs, as well as underground mains and traffic tunnels.

The injection strengthened and functionally renewed the brick foundations of the museum that had undergone application of the modern cement-polymer waterproofing materials produced by Schomburg GmbH & Co. KG (Germany) to preclude the possibility of future moistening. With the aim of significant prolongation of service life, the waterproofed foundation surfaces were additionally subject to covering with watertight polymer mats (Fig. 8.18 in Appendix).

The specialists of the SPC Center "Techno-Resurs" NASU fulfilled similar operations in the stone foundations and inside basements of the hotel "Leopolis" in Lviv (Fig. 8.22). Figure 8.23 in the Appendix shows the general appearance of the strengthened basements of the hotel and stone masonry joints protected by the modern cement-polymer waterproofing materials. Similar services are offered by many foreign firms.

Fig. 8.23 General appearance of the strengthened basements of the hotel "Leopolis" in Lviv after injecting and waterproofing operations

Fig. 8.24 Repaired water fountain with statue of Greek God of the Seas Neptune at Rynok Plaza in Lviv

Using the modern injection polyurethane and waterproofing cement-polymer materials, the specialists of the SPC Center "Techno-Resurs" NASU repaired four sculptural groups of Greek gods and structures of water fountains on the Rynok Plaza in Lviv (Fig. 8.24 in Appendix).

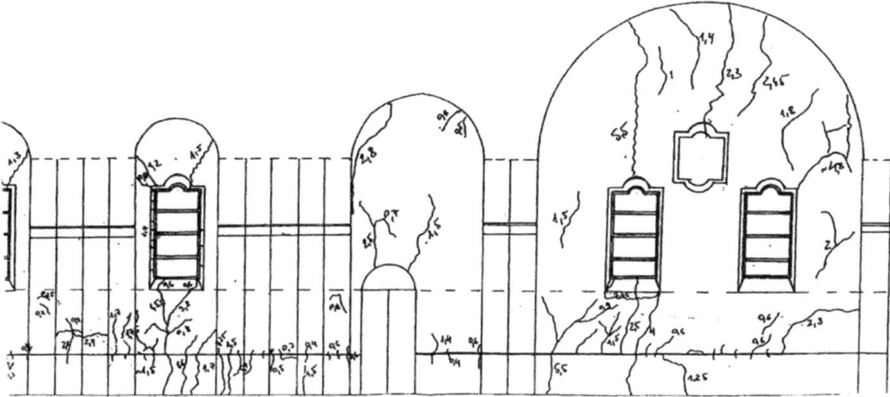

Fig. 8.25 Cracks in the walls and vault of the parish cathedral in Krzeszowice (Małopolska Province, Poland)

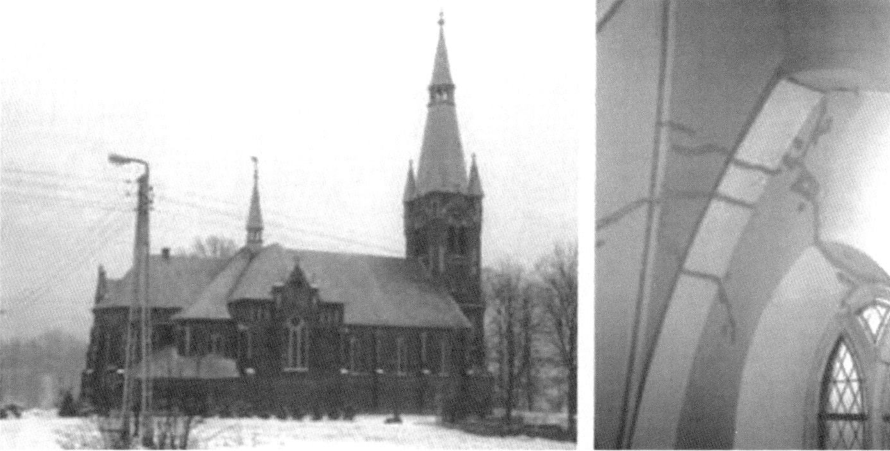

Fig. 8.26 General appearance and fragment of restored window structure of the parish cathedral in Krzeszowice. (Małopolska Province, Poland; photo courtesy of the company CarboTech-Polonia)

The injection strengthening technologies based on fluent polymer materials find especially wide application in the countries of Western and Eastern Europe (Germany, France, Czech Republic, Poland) that have an abundance of religious structures, palaces, antique bridges, etc. Figures 8.25 and 8.26 (the latter in the Appendix) demonstrate damages in the walls and vault of the parish cathedral in Krzeszowice (Małopolska Province, Poland), general appearance, and a fragment of the restored complicated window structure. The dangerous cracks and fractures in the walls and vault of the cathedral had been caused by systematic land subsidence due to operations of the neighboring underground coalmine.

Fig. 8.27 Antique bridge pillar in Germany

Bridges

Antique bridges create substantial technological difficulties for the healing of cracks and damages. While a bridge roadway is readily accessible for repair or replacement, the load-carrying capacity and reliability of stone or brick bridge pillars is subject to enhancement only through injection of fluent polymer compositions (Figs. 8.27, 8.28 in Appendix).

Using such injection technologies, the serviceability of concrete pillars and load-carrying elements was renewed for two road bridges over the rivers of Latoritsa near Svalyava town (Zakarpats'ka Oblast, Ukraine) and Limnitsa in Vistova village (Ivano-Frankivs'k region, Ukraine).

Renewal of Damaged Concrete and Reinforced Concrete Structures in the Illichevsk Seaport Using Cement-polymer Mortars

As shown in the previous sections, the prolonged effect of mechanical factors combined with permanent action of chemically reactive environments and temperature variations on concrete and reinforced concrete structures leads to impairment and

Fig. 8.28 Cracked stone masonry of antique bridge, inserted packers, and excess of strengthening polymer composition extruded through joints between stones. (photo courtesy of MC-Bauchemie Müller GmbH & Co)

the accelerated failure of these structures [4–6], [13]. The failure as a rule causes spalling of a concrete matrix surface and exposure of steel reinforcements, as well as formation and growth of deep or often through cracks and laminations.

The severe conditions of seaport use accelerate the stress corrosion processes in concrete and reinforced concrete structures. Specific mechanisms of damage depend on the long-term action of seawater, saline mist, sea slaps, ice floes, and abrasive wear by sand and other hard particles, as well as static and cyclic mechanical loads. In addition, the cement stone of concrete undergoes action of chemically and/or microbiologically aggressive media (polluted with industrial air emissions, wastewater, aqueous oily mixtures, etc.). The mechanical factors include the stresses trapped in the design during mounting of the reinforced concrete structures and/or arisen during in-service sagging of foundations, supports, or the structures themselves due to earth moving, sea slaps, strains and vibrations during use by vehicles, cyclic temperature changes, etc. [5].

The most effective way to eliminate the essential stress corrosion damages, beginning with the vast areas of surface defects and peeling and deep inner cracks, and restoration of serviceability of the structures immediately on site, consists of replacement of the impaired surface concrete layers with modern cement-polymer mortars with additional waterproofing and application of protective paint coatings (Figs. 8.29 and 8.30).

The technology of repair and restorative works for the damaged concrete and reinforced concrete structure of seaports comprises the following stages, materials, and equipment. The basic injection materials for the filling and strengthening of stress corrosion cracks, laminations, voids and other damages in large-size berth constructions are poured in water polymer-cement compositions ASOCRET made by Schomburg GmbH & Co. KG (Germany). Repair and grouting mortars ASOCRET tolerate preparation directly on site by means of introducing water into the dry polymer-cement powder and mixing it immediately before introduction into the concrete matrices [13].

Fig. 8.29 Striking a crack in a concrete matrix before filling with repairing aqueous cement-polymer mortar

Fig. 8.30 Application of a surface layer of cement-polymer material onto the filled crack and neighboring areas

During such serviceability renewal of structural elements, the process flow sheet includes the following stages or procedures:

- Visual inspection of damages in large-size berth constructions and instrumental measurements (presence, number, and type of cracks, laminations and/or other defects);
- Engineering and technical calculations of required repair compositions and material volumes;

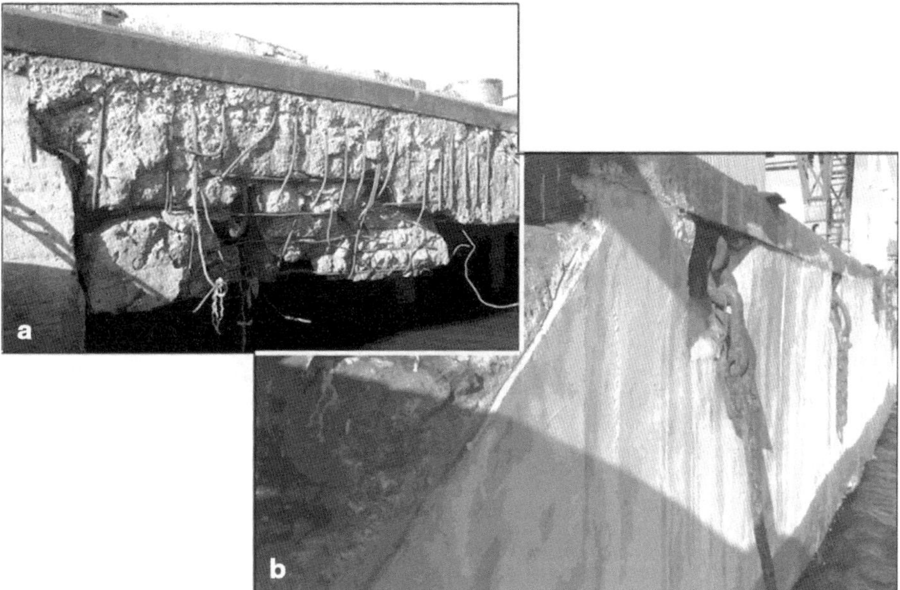

Fig. 8.31 Damaged (**a**) and repaired using cement-polymer materials (**b**) in a berth in the Illichevsk seaport

- Execution of effective technological operations for the most complete mechanical removal of surface and subsurface layers of 'weak' and loosely attached concrete around the cracks, laminations, and other defects as possible;
- Injection filling of the defects

The stage of cleaning (peeling or refining) the exposed surfaces of concrete and steel reinforcement before applying the polymer-cement compositions is very important for ensuring renewal of serviceability of the damaged structure. Such cleaning is obtainable using sandblasting until reaching the first surface finish class according to GOST 9.402-89 or equivalent finishing level Sa $2^1/_2$ according to ISO 8501-1. Optionally, a steel brush treatment can be used instead of sandblasting until reaching the second surface finish class according to GOST 9.402-89 or finishing levels St2 ... St3 according to ISO 8501-1. If needed, the cleaned surfaces can be additionally degreased using a woolen cloth wetted with an organic solvent (acetone, clear petrol, trichloroethylene, etc.) immediately before applying the polymer-cement compositions.

The finishing pre-starting procedure consists of applying two or three layers of the anticorrosion polymer-cement composition ASOCRET-KS/HB poured into water in a total amount of $1.6\,\text{kg/m}^2$ onto the cleaned and degreased surface of a steel reinforcement using paint rollers or hairy brushes. Surface priming of cleaned and degreased concrete around each damage or defect consists of applying one layer of the composition ASOCRET-KS/HB in a total amount of $2.0 \ldots 4.0\,\text{kg/m}^2$.

Fig. 8.32 A crack in a reinforced concrete slab of the berth

Fig. 8.33 Reinforced concrete slabs of the railroad bed in the Illichevsk seaport after repair using polymer-concrete mortars

Recommended defect-filling technology: Cracks or voids in concrete and reinforced concrete are filled using a hand spatula. The specialized polymer-cement materials are preferable for different cases. In particular, the best material for filling damages with a depth of 10 . . . 40 mm is ASOCRET-FM40 in an amount of 2.2 kg/m^2/mm of layer thickness whereas damages with a depth of 30 . . . 110 mm require ASOCRET-VM-K100 in the same amount per one mm of layer thickness.

After filling the cracks and voids in the concrete or reinforced concrete with the repair materials using a spatula, but before final hardening of these materials, the surfaces are subject to smoothing by means of hand rubbing using wooden or cellular polystyrene bars. Then, the smoothened and completely hardened surface of repair material and concrete is subject to putty with the surface layer of ASOCRET-FS with a thickness of 6 . . . 10 mm in an amount of 1.6 kg/m^2/mm of thickness. The final treatment of puttied reinforced concrete berths requires a special coating of paint to protect the constructions from carbonation and sea atmosphere.

For this purpose, the surface is subject to priming with the composition ASOCRET-OS/TG, painting with the composition ASOCRET-OS/BF, and finishing with the protective elastic composition ASOCRET-OS/RS. Figures 8.31–8.33 show the practical results of integrated works on renewal of damaged concrete and reinforced concrete structures using the modern polymer-cement compositions ASOCRET.

Each restorable structure in the seaport was subject to individual procedures of integrity and serviceability renewal using cement-polymer mortars taking into account its specific damaged status and performance. The structures with deep cracks, laminations and/or other defects underwent application of the cement-polymer mortars, as well as waterproofing coatings only after injection strengthening with polyurethane or polyepoxy compositions [14], [15].

References

1. Panasyuk VV, Sylovanyuk VP, Marukha VI (2005) Strength of cracked structure elements healed using injection technologies. Phys Chem Mech Mater 6:60–64
2. Marukha VI, Serednitskiy YA, Gnip IP, Sylovanyuk VP (2007) Development of injection technologies and design of mobile equipment for diagnostics and serviceability restoration of concrete or reinforced concrete structures operating under stress-corrosive conditions. Sci Innovat 3:55–62
3. Sylovanyuk VP, Marukha VI, Genega BY, Ivantyshyn MA (2002) Fracture mechanics as scientific basis of densification injection technology in long-term objects reconstruction. In: Mekhanika i fizika ruinuvannya budivel'nykh materialiv i konstruktsii (Fracture mechanics and physics of building materials and structures). Issue 5, Kamenyar, Lviv, pp 373–382
4. Marukha VI, Genega BY (2001) Sealing technologies for strengthening and repair of reinforced concrete structures. In: Diagnostika, dovgovechnost' ta rekonstruktsiya mostiv i budivel'nykh konstruktsiy (Diagnostics, durability, and reconstruction of bridges and concrete structure). No. 8. Kamenyar, Lviv, pp 158–161
5. Marukha A, Genega B, Serednitskiy Ya, Zaplatins'kiy M (2006) Concrete structure protection against stress corrosion using polyurethane injection compositions. Phys Chem Mech Mater 5:834–840
6. Marukha VI, Genega BY, Serednitskiy YA (2007) Technology of serviceability restoration using polyurethane injection compositions for concrete and reinforced concrete structures with stress-corrosion cracks. In: Scientific, resource, and technological potential realization efficiency in modern conditions. Proc. 7th Int. Ind. Conf., Lviv, Feb. 2007, pp 144–147
7. Sylovanyuk VP, Marukha VI, Genega BY, Onishchak NV (2007) Scientific basis of injection technologies for renewal damaged long-term structures. Ibid, pp 344–346
8. Czarnecki L, Emmons PH (2002) Naprava i ochrona konstrukcji betonowych (Repair and protection of concrete structures). Polski Cement, Krakov
9. Allen R, Edwards S (1993) Repair of concrete structures. Blackie Acad Prof, Glazgow
10. Marukha VI (2007) Strength of concretes with stress concentrators filled by polyurethane mixtures. In: Mekhanika i fizika ruinuvannya budivel'nykh materialiv i konstruktsii (Fracture mechanics and physics of building materials and structures). Issue 7, Kamenyar, Lviv, pp 524–534
11. Jankowski P (1993) Sealing of reinforced concrete structures using chemical injection. Eng Construct 3:101–104
12. Podolski B, Suwalski J, Wydra W (2000) The specifics of repair of old reinforced concrete building structures. Corros Protect 1:67–69

13. Czarnecki L, Skwara J (1998) Naprawa rys konstrurcji żelbetonowych metodą iniekcji (Injection repairing reinforced concrete structures with cracks). In: Proc. XIII All-Poland Conf. "Structural Designer Job Workshop", Ustron, Poland, pp 39–55
14. Scislewski Z (1999) Ochrona konstrukcji zelbetonowych (Protection of reinforced concrete structures). Arkady, Warsaw
15. Marukha VI, Vasilechko VO, Genega BY et al (2003) Waterproofing cover selection rules for protection of a sewage collector against very aggressive media volleys. In: Proc. int. water forum "Aqua Ukraine 2003", Ukrainian Water Association, Kyiv, Nov. 4–6, 2003, pp 194–195

Appendix

General Conclusions and Recommendations

1. Injection technologies are an effective means of renewal of damaged concrete and reinforced concrete objects of long-term operation in hydropower engineering, capital construction, and municipal services.
2. To restore the responsible objects serviceability it is very important to perform the timely technical diagnostics of the constructions state in order to detect early crack-like defects, which need in "healing" by injection technologies. For this purpose, the methods and systems described in Chap. 7 can be useful.
3. The choice of injection materials and specific features of technological operations of injection strengthening depend on the type and scope of damages in the object, its operation conditions and other factors (see Chap. 3 and 5).
4. Types and design of technological equipment for realization of injection technologies depend mainly on the nature and service characteristics of the used injection materials. Practical recommendations are in Chap. 4.
5. To realize injection technologies during repairing, the moving diagnostic-restoration complex mounted on a motor van is effective in many cases (see Chap. 4).
6. To perform diagnostic inspection of reinforced concrete sewage collectors and small-diameter water pipelines, a self-propelled diagnostic video device is useful that records and transmits to computer the size, character and location of damages in communications (Chap. 4).
7. Concrete and concrete-polymer mortars fill in the cracks and damages effectively if the distance between their surfaces (opening) is more than 1.0 mm, while polymer injection compositions are effective starting from 0.1 mm (Chap. 4).
8. The polyurethane compositions have the best technological and functional parameters among other injection polymer compositions including polyepoxy, silicon-organic, polyacrylic, and others. They are able to foam during interaction with water with a 10-fold volume increase. Practically this is the only injection material capable to stop pressurized water leakage or flow through cracks and damages in hydraulic constructions and structures (see Chap. 5).

V. V. Panasyuk et al., *Injection Technologies for the Repair of Damaged Concrete Structures,* DOI 10.1007/978-94-007-7908-2,
© Springer Science+Business Media Dordrecht 2014

9. If strength of adhesion between the host (concrete) and guest (injected) materials is close to cohesion strength of the concrete, the injection material hardness becomes the main parameter determining the injection effectiveness in the damaged object renewal (see Chap. 6).

10. Completeness (degree) of defect (e.g. crack) filling with an injection material is very important. If, for some reason, it is impossible to fill the defect along its full length, even its partial filling can significantly improve serviceability of the damaged structural element. In this case, it is very important to fill in the crack tips. Calculations (Chap. 6) show that filling only 10% of crack length can provide structural element renewal effect about 80% of maximum possible value (for a completely filled crack).

11. To optimize the conditions of crack injection in a deformed body, it is necessary to pre-calculate the pressure of the injection mixture supply in order to prevent the early crack growth before fixing its edges due to polymerization or crystallization of injection materials (Chap. 6).

12. The crack wedging effect due to injection material hardening is not crucial for concrete serviceability. The model of an elastic wedge described in Chap. 6 allows predicting and estimating this effect.

13. If a body contains a system of cracks rather than a single crack, it is important to stop the cracks merging into a main crack. For this purpose, it is necessary to inject the closer tips of neighboring cracks first (Chap. 6).

14. To restore objects with volume defects (voids), it is necessary to use injection materials with hardness as close as possible to the hardness of the host material. One of methods for improvement of injection material hardness consists in introduction of the hard filler particles into the polyurethane (see Chap. 6).

15. A complete renewal of structural element damaged by a plane (thin) crack is sometimes possible even at injection of the material with a significantly lower (in some cases by two orders of magnitude) hardness than that of the host material. This fact is very important in practice, since injection materials possessing a good penetrability usually have low hardness after hardening. These are for example polyurethanes. At the same time, the use of such injection materials provides a high level of renewal of damaged structural elements (see Chap. 6).

16. A drawback of conventional materials such as water-cement mortar or water-cement-sand mortar when used for filling cracks in building constructions consists in insufficient adhesion to concrete. Experimental studies with polyurethanes (as injection materials) and concretes (as host materials) made it possible to find the composition with adhesion strength exceeding the cohesive strength of concrete. This characteristic of injection material is one of the most important for the effective renewal of damaged long-term operation objects (see Chap. 6).

17. If large volumes of injection materials are required to fill in the large volumes of defects, the mixtures of polyurethane with hard fillers such as sand, wollastonite, and so forth are preferable to reduce production costs. It has been found that the specimens containing defects "healed" with such composite materials reveal strength at least the same as those injected by pure polyurethane.

Printed by Printforce, the Netherlands